Optofluidics 2015

Special Issue Editors

Shih-Kang Fan
Da-Jeng Yao
Yi-Chung Tung

MDPI • Basel • Beijing • Wuhan • Barcelona • Belgrade

MDPI

Special Issue Editors
Shih-Kang Fan
Department of Mechanical Engineering,
National Taiwan University
Taiwan

Da-Jeng Yao
Institute of NanoEngineering and MicroSystems,
National Tsing Hua University
Taiwan

Yi-Chung Tung
Research Center for Applied Sciences,
Academia Sinica
Taiwan

Editorial Office
MDPI AG
St. Alban-Anlage 66
Basel, Switzerland

This edition is a reprint of the Special Issue published online in the open access journal *Micromachines* (ISSN 2072-666X) from 2015–2016 (available at: http://www.mdpi.com/journal/micromachines/special_issues/optofluidics2015).

For citation purposes, cite each article independently as indicated on the article page online and as indicated below:

Author 1, Author 2. Article title. *Journal Name*. **Year**. Article number/page range.

First Edition 2017

ISBN 978-3-03842-468-0 (Pbk)
ISBN 978-3-03842-469-7 (PDF)

Table of Contents

About the Special Issue Editors

Shih-Kang Fan is currently a Professor in the Mechanical Engineering Department and a Researcher in the Center for Biotechnology, National Taiwan University (NTU). He received his B.S. from National Central University, Taiwan in 1996 and the M.S. and Ph.D. degrees from University of California, Los Angeles (UCLA) in 2001 and 2003, respectively. Dr. Fan started his academic career in 2004 and served as an Assistant Professor and then an Associate Professor in the Institute of Nanotechnology and the Department of Material Sciences, National Chiao Tung University, Taiwan. He moved to NTU in 2012. His research interests are in the areas of electrowetting, electromicrofluidics, tissue engineering, and in vitro diagnosis.

Da-Jeng Yao is currently a Professor and director at Institute of NanoEngineering and MicroSystems (NEMS), National Tsing Hua University, Taiwan. He is also the adjunct professor in Department of Power mechanical engineering, and Department of Engineering system and science, at the same university. He received his MS from Department of Mechanical Engineering, Lehigh University in 1996, and Ph.D. from Department of Mechanical and Aerospace Engineering, University of California at Los Angeles (UCLA) in 2001. His research interests are in the areas of bio-sensing system, neuron engineering, bio-sample preparation system and microfluidic reproductive medicine on a chip.

Yi-Chung Tung is currently an Associate Research Fellow at Research Center for Applied Sciences, Academia Sinica, Taiwan. He received his B.S. and M.S. degrees in Mechanical Engineering from National Taiwan University, in 1996 and 1998, respectively. In 2004 and 2005, he received his M.S.E. degree in Electrical Engineering and Ph. D. degree in Mechanical Engineering from University of Michigan. Before joining Research Center for Applied Sciences, Academia Sinica, he worked as a Postdoctoral Research Fellow at University of Michigan from 2006–2009. His research interests are in the areas of Integrated Biomedical Microdevices, Cell Culture in Various Micro-Environments, Micro/Nanofluidics, Polymer/Silicon Hybrid Microsystems, Advanced Micro/Nano Fabrication Techniques.

Preface to "Optofluidics 2015"

Optofluidics combines and integrates optics and fluidics to produce versatile systems that are difficult to achieve through either field alone. With the spatial and temporal control of the microfluids, the optical properties can be varied, providing highly flexible, tunable, and reconfigurable optical systems. Since the emergence of optofluidics, numerous systems with varied configurations have been developed and applied to imaging, light routing, bio-sensors, energy, and other fields. This Special Issue aims to collect high quality research papers, short communications, and review articles that focus on optofluidics, micro/nano technology, and related multidisciplinary emerging fields. The Special Issue will also publish selected papers from The Fifth International Conference on Optofluidics (Optofluidics 2015) held in Taipei, Taiwan, from July 26–29, 2015. Optofluidics 2015 covered the latest advances and the most innovative developments in micro/nanoscale science and technology. The aim of this conference was to promote scientific exchange and to establish networks between leading international researchers across various disciplines. Approximately 300 delegates participated in Optofuidics 2015 from across the globe. In total, 242 presentations were arranged, including 10 plenary speeches, 27 keynote speeches, 65 invited talks, 33 contributed talks and 107 poster presentations with topics ranged from fundamental research to its applications in chemistry, physics, biology, materials and medicine. This Special Issue "Optofluidics 2015" is a collection of ten papers on this interdisciplinary research field.

Several important optofluidic components are collected in this Special Issue. Le et al. [1] investigated the performance of liquid-core/liquid-cladding microlenses for tunable in-plane beam focusing. Testa et al. [2] reviewed the fabrications, characterizations, and sensing applications of liquid core antiresonant reflecting optical waveguide (ARROW). In addition, an optofluidic Fabry–Pérot micro cavity was developed by Gaber et al. [3] as an optical sensor for liquid refractometry. Various optofluidic immunosesors were reviewed by He et al. [4] for point-of-care diagnosis including colorimetric and plasmonic mechanisms. Lu et al. [5] reported a two-layer microstructure fabrication technique through a single step anisotropic wet etching with residue deposition for possible plasmonic applications.

This Special Issue also collects several papers regarding cell studies. Zhang et al. [6] reviewed the microflow cytometers with various flow focusing and light collection methods. Matteucci et al. [7] described the fabrication and assembly of an injection-molded optofluidic system with integrated optical fibers for the application of optical stretching and Raman spectroscopy. The principle—application of cell sorting and drug treatment, and heating of optical stretching for cell mechanical property characterization—was elaborated and reviewed by Yang et al. [8] An alternative scheme to evaluate the single cell mechanical property with constriction channels was reviewed by Xue et al. [9] Moreover, Tsao et al. [10] examined the rat bone marrow stromal cell differentiation with the stimulation of mechanical shear stress under varied flow rates.

We express our gratitude for the financial support received from Ministry of Science and Technology (Taiwan), Bureau of Foreign Trade (Taiwan), National Taiwan University and Research Center for Applied Sciences of Academia Sinica and for the administrative support received from Instrument Technology Research Center in making Optofluidics 2015 a successful conference. Our acknowledgements include Nam-Trung Nguyen, Mengdie Hu and all staff from Micromachines for their kind assistance during the preparation, and, most importantly, all authors who have contributed their work to this Special Issue.

Shih-Kang Fan, Da-Jeng Yao and Yi-Chung Tung

Special Issue Editors

References

1. Zichun Le, Yunli Sun and Ying Du, Micromachines 2015, 6(12), 1984–1995.
2. Genni Testa, Gianluca Persichetti and Romeo Bernini, Micromachines 2016, 7(3), 47.
3. Noha Gaber, Yasser M. Sabry, Frédéric Marty and Tarik Bourouina, Micromachines 2016, 7(4), 62.

4. Jie-Long He, Da-Shin Wang and Shih-Kang Fan, Micromachines 2016, 7(2), 29.
5. Han Lu, Hua Zhang, Mingliang Jin, Tao He, Guofu Zhou and Lingling Shui, Micromachines 2016, 7(2), 19.
6. Yushan Zhang, Benjamin R. Watts, Tianyi Guo, Zhiyi Zhang, Changqing Xu and Qiyin Fang, Micromachines 2016, 7(4), 70.
7. Marco Matteucci, Marco Triches, Giovanni Nava, Anders Kristensen, Mark R. Pollard, Kirstine Berg-Sørensen and Rafael J. Taboryski, Micromachines 2015, 6(12), 1971–1983.
8. Tie Yang, Francesca Bragheri and Paolo Minzioni, Micromachines 2016, 7(5), 90.
9. Chengcheng Xue, Junbo Wang, Yang Zhao, Deyong Chen, Wentao Yue and Jian Chen, Micromachines 2015, 6(11), 1794–1804.
10. Chia-Wen Tsao, Yu-Che Cheng and Jhih-Hao Cheng, Micromachines 2015, 6(12), 1996–2009.

micromachines

MDPI

Article

Optofluidic Fabry-Pérot Micro-Cavities Comprising Curved Surfaces for Homogeneous Liquid Refractometry—Design, Simulation, and Experimental Performance Assessment

Noha Gaber [1,2,*], Yasser M. Sabry [3], Frédéric Marty [1] and Tarik Bourouina [1]

[1]   Laboratoire Electronique, Systèmes de Communication et Microsystèmes, Université Paris-Est, ESIEE Paris, ESYCOM EA 2552, 93162 Noisy-le-Grand, France; frederic.marty@esiee.fr (F.M.); tarik.bourouina@esiee.fr (T.B.)
[2]   Center for Nanotechnology, Zewail City of Science and Technology, Sheikh Zayed District, 6th of October City, 12588 Giza, Egypt
[3]   Electronics and Electrical Communication Engineering, Faculty of Engineering, Ain-Shams University, 1 Elsarayat St., Abbassia, Cairo 11517, Egypt; yasser.sabry@eng.asu.edu.eg
*   Correspondence: ngaber@zewailcity.edu.eg or noha.gaber@esiee.fr; Tel.: +20-23-854-0476

Academic Editors: Shih-Kang Fan, Da-Jeng Yao and Yi-Chung Tung
Received: 28 February 2016; Accepted: 1 April 2016; Published: 7 April 2016

Abstract: In the scope of miniaturized optical sensors for liquid refractometry, this work details the design, numerical simulation, and experimental characterization of a Fabry-Pérot resonator consisting of two deeply-etched silicon cylindrical mirrors with a micro-tube in between holding the liquid analyte under study. The curved surfaces of the tube and the cylindrical mirrors provide three-dimensional light confinement and enable achieving stability for the cavity illuminated by a Gaussian beam input. The resonant optofluidic cavity attains a high-quality factor (Q)—over 2800—which is necessary for a sensitive refractometer, not only by providing a sharp interference spectrum peak that enables accurate tracing of the peak wavelengths shifts, but also by providing steep side peaks, which enables detection of refractive index changes by power level variations when operating at a fixed wavelength. The latter method can achieve refractometry without the need for spectroscopy tools, provided certain criteria explained in the details are met. By experimentally measuring mixtures of acetone-toluene with different ratios, refractive index variations of $0.0005 < \Delta n < 0.0022$ could be detected, with sensitivity as high as 5500 μW/RIU.

Keywords: Fabry-Pérot cavity; optical resonator; optofluidic sensor; on-chip refractometer; refractive index measurement; lab-on-a-chip

---

1. Introduction

Refractometry is a well-known optical characterization technique to identify dielectric materials by measuring their refractive index (RI). It has various applications in chemistry, environmental science, material science, and even biology, as the RI of cells provides an important insight about its properties [1]. There are widely vast refractometry techniques depending on various optical phenomena such as beam refraction [2], light wave interference [3], optical cavity resonance of different configurations [4,5], and surface plasmon resonance [6]. Some of these techniques are matured and form the core of several products available in the market. Benchtop Abbe refractometers, for instance, are important in many labs whether they are optics labs, chemistry labs, or even material science labs. However, with the trend of miniaturizing devices, several on-chip versions of refractometers have emerged to provide cheaper and more compact devices that require small sample quantities.

Many attempts has been made to integrate various refractometry techniques on chip, with each having their advantages and drawbacks. To ease the comparison, these techniques can be categorized under two major themes: "surface refractometry" and "volume refractometry" according to the amount of light and the nature of the light waves that pass through and interact with the sample material for sensing it. In surface refractometry, the sensing mechanism employs only the evanescent part of an electromagnetic wave to interact with the sample located at the surface of the resonator. Although these techniques can reach very high resolution of detecting RI change Δn up to 9×10^{-9} RIU [7], they are generally at risk of being affected by surface contamination and not suitable for applications requiring thick surface penetration like measuring through big biological cells. For instance, the microtube ring resonators [8] can be used as optofluidic refractometric devices [9] with Q-factors up to several thousands [10] and sensitivities up to 880 nm/RIU for passive resonators [11], or even 5930 nm/RIU for an optofluidic tube coupled with a ring laser [12]. However, their evanescent field can interact with the surrounding environment at the distance of only few hundred nanometers inside and outside the microtube [11]. So imagine the application of water contamination detection as an example, the impurities will not be detected unless they accidentally swim beside the resonator surface through its small evanescent field area. On the contrary, in "volume refractometry" techniques, the light wave totally propagates through the sample renders the depth of interaction greatly increased. So, the sample can be entirely scanned by the detecting light beam when necessary. This comes at the expense of the attained sensitivity and resolution that is generally lower than the surface refractometry counterparts. Then it is useful to carefully choose the detection technique upon the priority required in the intended application.

Amongst the volume refractometry techniques to measure the RI of liquids, is the Fabry-Pérot (FP) optical resonator. It is simply a cavity of length (d) enclosed between two mirrors that causes the light to be multiply reflected between these mirrors. The length d is designed to provide constructive interference at certain wavelengths (λ), as it is equal to multiples of $\lambda/2$, leading to maximum signal at the output of the resonator. For other wavelengths, destructive interference with various degrees occurs, giving different signal levels with less values; hence, an interference spectrum is obtained for different incident wavelengths. By introducing the liquid sample inside the cavity, the peaks of the spectrum shift according to the RI of this liquid (n) and the corresponding effective cavity length ($n \cdot d$), enabling the detection of the RI. The conventional detection method is done by recording the spectrum for at least one interference peak for the sample and for a reference liquid, then measure the shift between them. This method requires expensive equipment such as an optical spectrum analyzer or tunable laser source. An alternative method depending on the signal level variation has been introduced, first with ring resonators [3], then with FP resonators by our group [5]. The goal of this method is eliminating the need for such expensive spectrometry equipment, as only a photodetector is required to read the power change at a single wavelength. This can be achieved with enough sensitivity only if relatively high values of quality factors are attained by the resonator. For ordinary FP microfluidic refractometer with flat mirrors [13–15], Q-factor values are limited due to the light diffraction outside the cavity from the mirror boundaries as the mirror surface profile shape is not compatible with the Gaussian beam wave front's curved shape. By employing curved mirrors and a micro-tube as in our device, the beam can be well confined leading to high Q-factors. Thereby, the interference peaks will have fast roll off enabling high sensitivity demonstrated by our previous work [5]. However, in our previous attempts the analytes used had different attenuation values at the employed wavelength range, which necessitates recording the spectra and normalizing them before tracing the power change with the RI. In this paper, we show performance with toluene and acetone mixture as both these liquids interestingly have the same attenuation values at our wavelength band, which enables studying the refractometer performance to investigate the possibility of spectrum-free detection. The experimental part is preceded by detailing the device design using a developed model and numerical simulation by HFSS (High Frequency Structural Simulator) that help the assessment of the refractometer performance.

2. Design

For a highly sensitive refractometer, the FP resonators should provide sharp (or narrow) spectral lines to be more selective. That is represented by having a high Q-factor value at a specific wavelength. The Q-factor is given by:

$$Q = F\frac{2nd}{\lambda} \qquad (1)$$

where F is the finesse of the optical cavity that is defined as:

$$F = 2\pi N \qquad (2)$$

where N is the number of round trips after which the energy bouncing inside the cavity drops to 1/e of its starting value. Hence, minimizing the energy losses inside the cavity is essential. The diffraction effect can be a major source of energy loss, as the energy escapes out of the open cavity one round trip after another. It can be overcome by making a proper design for the cavity mirrors resulting in a stable cavity. Long cavities and large mirror size are preferable for the high Q-factor they can produce, but this is obviously not compatible with miniaturization. Indeed, a cavity based on small mirrors imposes using small spot size for the light beams, which causes excessive beam expansion and light escapes the open cavity, when planar mirrors are used. That is why the Q-factor was limited in the previous attempts found in literature that implemented on-chip FP refractometer with flat mirrors [13–15]. On the other hand, earlier reports about FP cavities with spherical mirrors have demonstrated their excellent focusing capability. The curvature of the mirror focuses the beam in both directions perpendicular to the beam direction of propagation. Despite their high performances, the spherical resonators are difficult to miniaturize practically since the standard micro-fabrication technologies allow the realization of in-plane curved surfaces only. Thereby, as an intermediate solution, cylindrical mirrors can be implemented though the micro-fabrication process. These in-plane curved cylindrical silicon mirrors are adopted to achieve partial confinement in one lateral direction only. To evaluate the performance of a Fabry-Pérot cavity realized by different mirror shapes, a model for the theoretical Q-factor has been developed using Equations (1) and (2), where the corresponding round trip number N for the finesse calculation is deduced from the equation:

$$|r_{m1}r_{m2}|^{2N} \prod_{n=1}^{N}\eta_n = e^{-1} \qquad (3)$$

where r_{m1} and r_{m2} are the field reflection coefficients of the cavity mirrors and η_n is the round-trip coupling efficiency between the output field from the cavity and the fundamental mode of the optical fiber used for injecting the light into the cavity. The value of the finesse is limited by the coupling loss and the mirrors reflection, as N is calculated by solving Equation (3). Detailed analysis of the coupling efficiency can be found in references [16,17] where this modelling methodology was successfully applied in designing optical cavity based on planar mirror facing a three-dimensional curved mirror exhibition microscale size and curvature. In our case, we are using the in-plane curvature of the deeply-etched cylindrical mirrors and the out-of-plane curvature of the fluidic micro-tube to achieve three-dimensional control of the diffraction effect and arrive at a stable optical cavity. The Q-factor of the FP cavity is analyzed in Figure 1 for the 2-D confinement achieved by cylindrical mirrors, and 3-D confinement achieved by spherical mirrors and compared to the flat surfaces case, where there is no control on the diffraction effect. The Gaussian beam waist radius of the input/output lensed fibers used for light injection/collection is $w_0 = 8$ µm and the input/output fiber tip location is close to input/output mirrors while the beam waist location is one Rayleigh away from the fiber tip. The power reflectivity of the mirrors is assumed 97% and its radius of curvature is set by 140 µm. The Q-factor is plotted *versus* the optical diffraction length, given by the physical propagation length divided by the refractive index, normalized to the light wavelength.

As deduced from Figure 1, the performance of a cavity with cylindrical mirrors is slightly better than that with flat for intermediate cavity length. But with spherical mirrors, the Q-factor values are superior. But stile, such 3-D mirror curvature cannot be easily fabricated on chip. A workaround to overcome that is to decouple the 3-D curvature into two surfaces. One surface for the in-plan direction, which is the cylindrical mirrors, and a cylindrical rod lens laying on the chip provides confinement in the out-of-plane direction. The rod lens is formed by a micro-tube with the analyte inside, serves simultaneously for delivering the liquid under test and for light confinement. The schematic of the employed devise is shown in Figure 2.

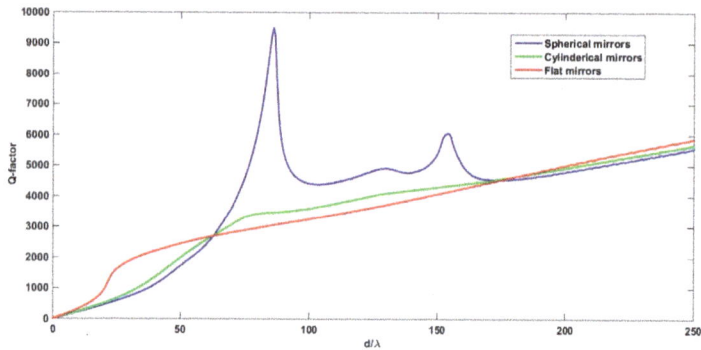

Figure 1. The theoretical values of the *Q*-factors for a FP cavity formed by straight mirrors, cylindrical mirrors, and spherical mirrors, plotted *versus* the diffraction length of the cavity.

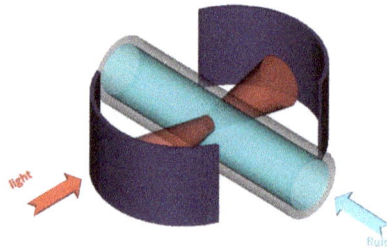

Figure 2. Schematic diagram of the cylindrical Fabry-Pérot cavity with the micro-tube inside.

For using the FP structure with the micro-tube in liquid analysis, the refractive index of the analyte may affect the cavity stability. Hence, the range of refractive indices that doesn't degrade the performance much should be determined. The stability of the FP cavity can be investigated in a simple way by the ray matrix approach [18]. In this approach, each encountered surface is represented by a matrix, and then the equivalent matrix is obtained by multiplying them either symbolically or numerically by Matlab. Let the equivalent matrix components be A, and B, in the first raw; C, and D, in the second row. Then the stability condition is having the stability parameter (A + D)/2 be less than or equal to 1. We assume that the light behavior is decoupled in *XZ* (horizontal) and *YZ* (vertical) planes. Hence, each cross section is treated as a 1-D problem with schematics shown in Figure 3.

Thereby, we have two conditions that should be met simultaneously to achieve full stability in both directions. These conditions are:

$$0 \leqslant \frac{2}{r_1}\left(2d_{air} + \frac{2d_s}{n_s} + \frac{d_t}{n_t}\right) - \frac{4d_{air}}{r_1^2}\left(d_{air} + \frac{2d_s}{n_s} + \frac{d_t}{n_t}\right) - \frac{1}{r_1^2}\left(\left(\frac{d_t}{n_t}\right)^2 + 4\frac{d_s}{n_s}\left(\frac{d_s}{n_s} + \frac{d_t}{n_t}\right)\right) \leqslant 1 \quad (4)$$

$$0 \leqslant 4\frac{d_{air}^2}{n_s^2}\left(\frac{n_s-n_t}{r_{tt}n_t}+\frac{1-n_s}{r_{ss}}\right)^2+\frac{4}{r_{tt}}\frac{(n_s-n_t)}{n_s^2 n_t}\left(4d_{air}+2r_{ss}-n_s\left(3d_{air}+r_{ss}\right)\right)$$
$$+4\frac{d_{air}}{r_{ss}}\left(\frac{2}{n_s^2}-\frac{3}{n_s}+1\right)+4\frac{r_{ss}}{r_{tt}^2}\left(\frac{n_s-n_t}{n_s^2 n_t}\right)^2\left(2d_{air}+r_{ss}\right)+4\frac{1-n_s}{n_s^2}+1 \leqslant 1 \tag{5}$$

where the geometrical parameters and refractive indices are shown in Figure 3. The parameters of the real device we have implemented are $d_{air} = 76$ μm and $r_1 = 140$ μm, and a fused silica capillary tube with dimensions of $d_t = 75$ μm and $d_s = 26$ μm. The stability may or may not be guaranteed according to the refractive index of the fluid inside the tube. Calculating the stability parameter for a range of refractive indexes from 1 to 2 to cover the condition of air and the majority of fluids that can be introduced inside the tube, as indicated in Figure 4, the stability is always assured in the horizontal plane. However, the vertical plane restricts it to the liquids whose refractive indexes are between 1.1526 and 1.6673. The proposed range of indexes constrains the applications of such device to some liquids only, which means gases are excluded as their refractive index is close to 1, if a high Q-factor is to be exploited.

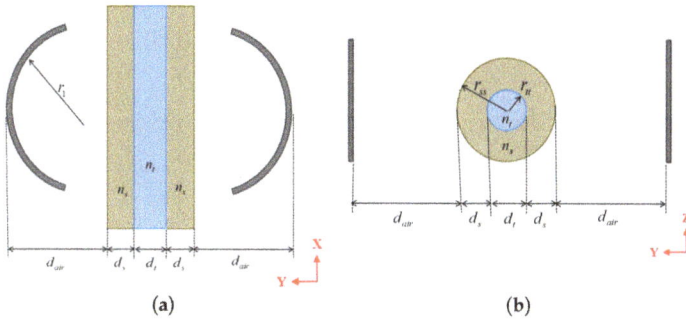

Figure 3. Schematic diagram for: (**a**) the horizontal cross section and (**b**) the vertical cross section, of the cylindrical Fabry-Pérot cavity with the micro tube inside indicating the design parameters and geometry.

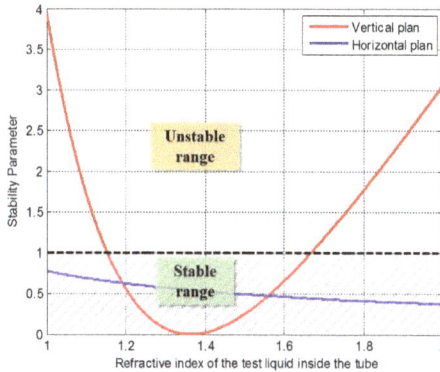

Figure 4. Stability parameter for different fluids inside the tube.

3. Simulation

For accurate representation for the light wave, a cavity with scaled-down dimensions is simulated by HFSS program based on finite element method to have an idea of the electromagnetic modes behavior inside such cavities. If cavities with real dimensions were to be simulated, enormous calculation resources would be required. To overcome this problem, miniaturized versions of the cavities have been designed and simulated. Moreover, to render the simulation more efficient, we exploited the symmetries of the design in respect to the XY and the YZ planes to simulate only one quarter the cavity volume. For further simplification and size reduction, cavities with a single silicon Bragg layer per mirror have been simulated. Also, the thickness of the silicon layer is taken equal to 111.4 nm equivalent to only one quarter of the wavelength (in silicon) with respect to the reference central wavelength of 1550 nm in vacuum. The scaled-down cavity has geometrical parameters of 9.85 μm for the physical length, 7.5 μm for the radius of curvature, 6.25 μm for the micro-tube eternal diameter, and 0.75 μm for the internal diameter; the spot size of the exciting Gaussian beam is 0.9 μm. Checking the stability of such downscaled cavity, the range of the filling liquid n_t that achieves stability in this case is between 1.15 and 2.03. Thereby we simulate a tube filed with different fluids that have different indices near the limits of that RI values range that achieves stability, lower and higher than the silica refractive index (the material of the walls of the tube). The selected values of the test fluid n_t are 1.18, 1.3, 1.6, and 1.8, all within the stability range. The transmitted output power spectra for these cases are shown in Figure 5.

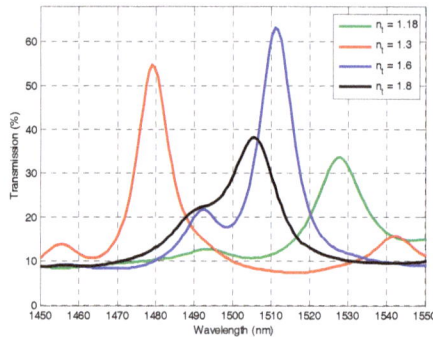

Figure 5. The transmission spectra of the curved cavity with a micro-tube filled with a test liquid of different refractive indices n_t.

As noticed from the transmission spectra, there is a large peak for each case along with one or several smaller peaks. The largest one corresponds to the fundamental resonance mode as revealed form the field distributions shown in Figure 6. While the smaller peaks correspond to the higher order Hermite-Gaussian transverse modes, as noticed from the field distribution for the higher order resonance mode at 1542 nm for the test liquid n_t = 1.3 shown in Figure 7. The cross sections have three spots in the transverse direction (note that only a quarter of the cavity is shown, hence we can see one and a half spot in X-direction, but it is mirrored around the Z-axis as we have even symmetry). These modes are typical resonance modes in FP resonators with spherical mirrors [19], and they appear also in FP cavities with cylindrical mirrors [20].

To investigate the field confinement quantitatively, Table 1 states the Q-factor values the confinement distances that is taken as the lateral distance from the maximum field value at the center of the spot to the value of half the maximum from the field distribution plotted Figure 6.

As theoretically predicted and as can be inherited from the field distributions in Figure 6, when the test fluid has refractive index less than that of silica, this may cause divergence of the beam after it refract at the internal surface of the micro-tube, which is the silica/test liquid interface. On the other hand, the test fluid with RI higher than that of the silica helps in confining the beam better and increasing the Q-factor. This happens until a certain extent, as the interspacing between the fundamental and higher order modes decreases with increasing the RI. When the interspacing is not enough to separate different peaks, intermodal interference occurs, like what is observed in the case of

$n_t = 1.8$ of Figure 5; in which, a strong coupling between the main peak and the side peak appears in the spectral response, and hence the main peak is smaller and wider; which renders the field spots to be of less intensity as in Figure 6d. The best performance regarding high transmission at the main peak, well confinement of light, and quite high Q-factor, is obtained with the case $n_t = 1.6$, which is larger than the refractive index of the tube material, but not too large to reduce the separation between modes much, causing their coupling; it is also away from the critical values of the stability conditions.

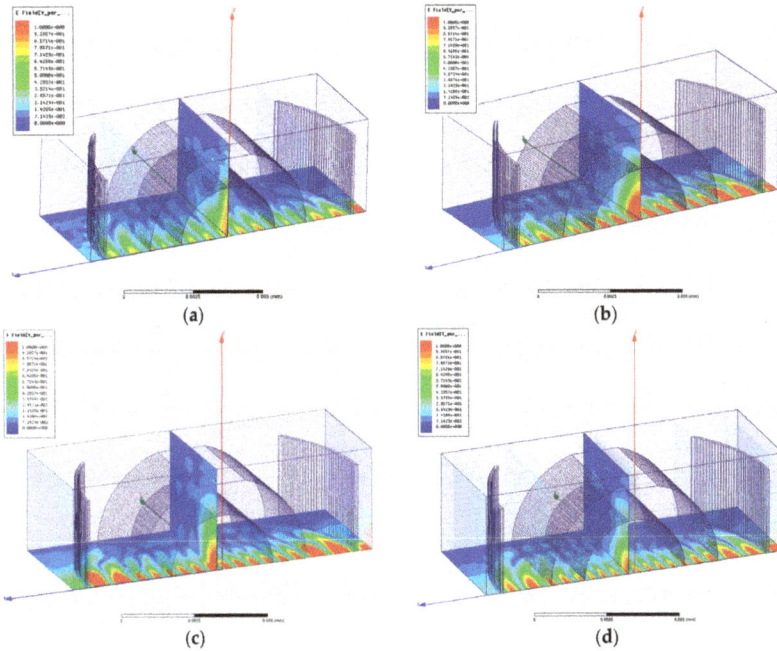

Figure 6. The electric field distribution at resonance for different test liquids (**a**) $n_t = 1.18$, resonance at 1528 nm; (**b**) $n_t = 1.3$, resonance at 1576 nm; (**c**) $n_t = 1.6$, resonance at 1511 nm; (**d**) $n_t = 1.8$, resonance at 1505.7 nm.

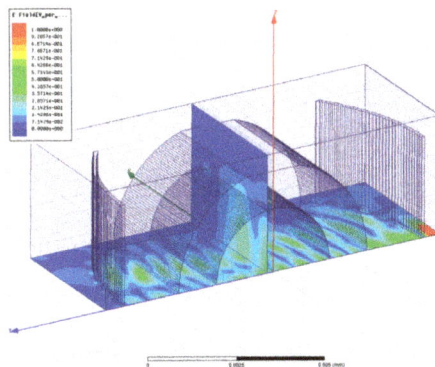

Figure 7. The electric field distribution of higher order resonance mode at 1542 nm for the test liquid $n_t = 1.3$.

A better quantitative comparison for the Q-factors and the confinement distances between the different cases is indicated in Table 1, from which one can notice that the confinement distance is smaller as n_t increases, which is predicted. However, for Q-factors, the trend is not the straight forward. The decreasing intermodal spacing between the main modes and the side ones causes interference between them, leading to reduction in the Q-factor, as most pronounced for $n_t = 1.8$.

Table 1. Comparison between the Q-factor and the confinement distance between different test liquid filling the tube.

RI of Test Liquid	$n_t = 1.18$	$n_t = 1.3$	$n_t = 1.6$	$n_t = 1.8$
Q_{peak}	70	129.5	128	55
Confinement distance	1.88 µm	1.34 µm	1.24 µm	1.11 µm

4. Materials and Methods

Figure 8a shows a cross section schematic of the device. The implementation of the cavity is realized by Deep Reactive Ion Etching (DRIE) process on silicon substrate. The etching of two Bragg mirrors spaced by 280 µm is done after transferring the pattern onto a 400 nm-thick thermal oxide layer, as a hard mask, through a lithography step followed by fluorinated plasma etching of the oxide. Each Bragg mirror consists of three layers of silicon/air pair with thicknesses 3.67 µm and 3.49 µm respectively; both thicknesses correspond to odd multiple of quarter the central wavelength—which is 1550 nm—in the two mediums. The channel depths are 80 µm measured by our 3D profile-meter. Figure 8b shows a Scanning Electron Microscope (SEM) image for the top part of one Bragg mirror that is indicated in the schematic by the dashed rectangle. It can be noticed that the scalloping effect associated with the DRIE process is noticeable only within less than 7 µm depth from the silicon surface, then the scalloping attenuates in deep regions due to the phenomenon known as Aspect Ratio Dependent Scalloping Attenuation (ARDSA) [21]. This phenomenon appears in narrow openings with high aspect ratios, and is attributed to the transportation limit of radicals [21]. The important depth to us that is illuminated by the light, is estimated between 15 and 20 µm. The SEM image in Figure 8c shows that this region is scallop-free within the intermediate walls, which is very beneficial to reduce the scattering light loss. Nevertheless, the outward surfaces of the first and last mirrors suffer from surface roughness as they are not confined by narrow trenches. Figure 8d shows noticeable scalloping with undercut length and etched depth per cycle estimated by 125 and 645 nm, respectively. The mentioned undercut value render the root mean square roughness (σ) for these surfaces equal to about 44 nm. This value can be considered much less than the used light wavelength; besides, only two surfaces from five have this value while σ is equal to almost zero for the other three surfaces. Thus we believe the scattering losses can be negligible in our case. The mirror verticality has been measured to deviating from the exact perpendicular angle by about 2° only, which has a negligible effect on the mirror performance. The pronounced effect was due to the wall thickness change as for slight fabrication error. Several measurements has be done for silicon walls thickness and the air gabs in-between from the SEM images. The most deviated measurements have been found to be 3.36 and 4.497 µm for the silicon walls and air gaps, respectively. That leads to a theoretically estimated maximum reduction in the mirror reflectivity from 99.74% at the designed dimensions to 88.56% at the measure ones.

After the silicon chip fabrication, the fused silica micro-capillary tube is inserted inside the cavity and connected to external tubing allowing for the liquid insertion. The interfaces of the micro-capillary are expected to introduce parasitic reflection loss due to refractive index change. Estimated by Fresnel formula, each interface with air causes about 3.3% reduction in the transmitted power. The inner interface of the tube will introduce some losses also, depending on the refractive index of the passing liquid, but it is expected to be even smaller as the refractive index difference will be smaller in case of a liquid than that with air. Different mixtures of toluene and acetone are used to perform the test since the absorption of both being almost the same. To insure that, spectroscopy of pure toluene and that of different mixing ratios with acetone is performed using IR-Affinity-1 Fourier transform infrared spectrophotometer from Shimadzu connected in transmission mode.

The optical testing setup for the chip consists of a tunable laser source of model 81949A and a detector head with a power meter of model 81634B from Agilent (Santa Clara Valley, CA, USA), controlled using a computer. Injecting and collecting the light into and from the cavity is done using lensed fibers from Corning with typical spot size of 18 ± 2 µm and 300 µm working distance. Fiber positioners of five degrees of freedom are used to align the fibers inside the input and output grooves on the chip.

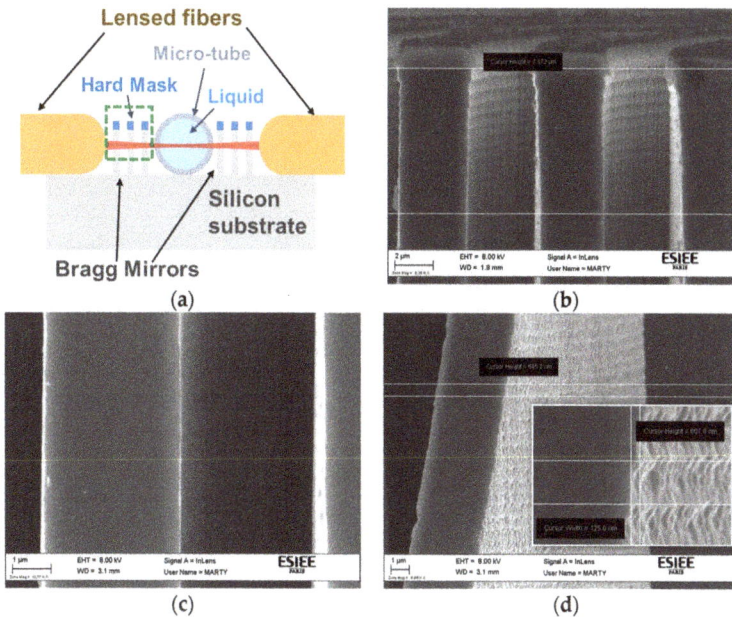

Figure 8. (a) Schematic diagram for the horizontal cross section of the device. (b) SEM image for the top part of one Bragg mirror indicated in schematic (a) by the dashed rectangle indicates the ARDSA phenomenon. (c) SEM image for the Bragg mirror at depth between 15 and 20 µm, corresponds to region illuminated by the light beams from the fiber, indicates the attenuatted scalloping. (d) SEM image for the outer surface of the Bragg mirror wall. The inset is a zoom indicating the dimensions of the non-attenuated scalloping.

5. Results and Discussion

The transmission from the cavity is recorded while filling the capillary by mixtures of toluene and acetone of different volumetric ratios. Figure 9 plots these transmission spectra simultaneously to allow comparison. As inherited from Figure 9a, the peaks between the wavelengths of 1588 nm and 1600 nm (surrounded by the circle) have the same power transmission values, despite the different mixing ratios of toluene and acetone. The discrepancy between the maximal power values of the different curves at this peak is found to be less than 0.53 µW, which may be attributed to slight temperature changes, laser power instability, or alignment disturbance upon changing the liquids and running the scan.

From Figure 9a, it can be noticed that the peaks do not all behave in the same way, some have decreasing levels upon increasing the concentration of toluene like those around the wavelength

1592 nm (indicated by the tangent dashed black line with negative slope); other peaks have decreasing levels like those between the wavelengths of 1594 and 1596 nm (indicated by the tangent dotted red line with positive slope). To investigate the reason behind that, a close look is needed at the two extreme curves (100% and 96.74% concentration of toluene), which are magnified around the wavelength 1595 nm in Figure 9b. One notice that the side peak of the higher order mode in the black curve (at wavelength of 1597 nm for the 100% toluene) is merged with the main peak for the red curve of the lower toluene concentration as the RI changes. This is apparently the reason behind reducing the transmission level of the later spectrum. Note that, similar behavior was numerically demonstrated with HFSS simulations in Section 3 as the refractive index test liquid changes. Note also that the Q-factor for the main peaks free from modal interference—such as the first peaks between 1587 and 1590 nm in Figure 9a—is the highest and can reach up to 2896, while it is less in other peaks due to coupling with the higher order resonance peaks, similar to what has been observed with numerical modeling.

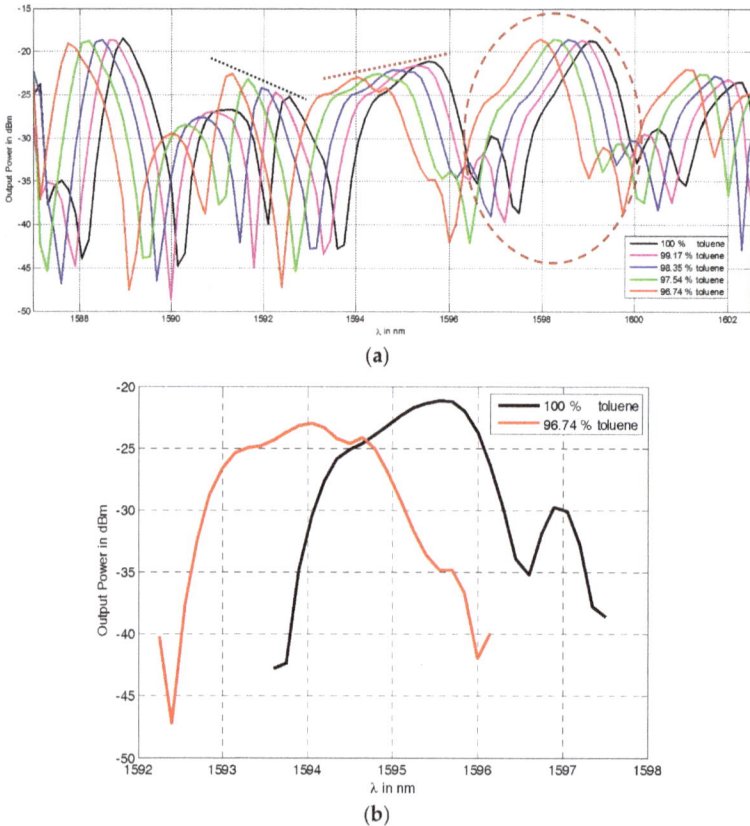

(a)

(b)

Figure 9. (a) The spectra of different mixture ratios of toluene and acetone measured by the proposed refractometry device. The peaks surrounded by the red dashed circle have the same maximum power transmission values, despite the different mixing ratios of toluene and acetone. (b) Zoom around the wavelength 1595 nm of the spectra of the two extreme mixing cases to indicate that the decrease in the transmitted power level upon changing the RI is due to the modal interference between the main peak and that of the higher order mode.

To characterize the performance of the refractometer, Figure 10 shows the zoomed view of the output power in μW *versus* wavelength in nm, which gives better linearity, around the selected peak. A reference line from the peak of a fitting curve of the pure toluene spectrum is used to trace the power drop upon the spectrum shift with the liquid RI changing. The error in wavelength between the measured peak and an interpolation is found to be less than 0.7 nm for all the curves and it is due to the poor measurement wavelength step of 1.5 nm.

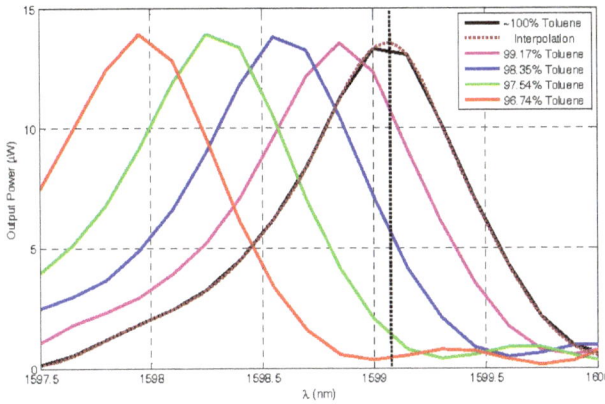

Figure 10. Zooming of the output power in μW *versus* wavelength in nm around the selected peak for refractometry analysis.

Due to the resonance peak shifting upon RI change, the power drops along the reference line at this wavelength. Hence, the refractometer can be performing by tracing the power value only at a single wavelength after calibrating the system only once, but the range of detection by this method is limited by the range at which the side of the resonance peaks are almost linear, so the last curve is excluded as it cuts the reference line outside the linearity region.

The common measurement technique by tracing the peaks' wavelength maxima is also performed. The wavelengths shift is plotted *versus* the toluene concentration in Figure 11. As the measurement wavelength step is 0.15 nm, the actual maxima wavelength may be allocated within error of ± 0.075 nm from the depicted values; this margin is indicated by the error bars in Figure 11. Note that the resulting relation is more like quadratic rather than linear, which indicates that the RI property for these liquids mixtures is not linearly additive upon volumetric ratios; hence it cannot be estimated from the known RI values of pure acetone and pure toluene. Similar nonlinearity has been observed for different liquid mixtures, which is thought to be caused be volume change upon mixing [22].

For the conventional method of tracing the peaks' wavelength shift, the sensitivity $\delta\lambda/\delta n_t$ is analytically deduced to be:

$$\frac{\delta\lambda}{\delta n_t} = \lambda\frac{d_t}{d} \tag{6}$$

The calculated sensitivity from the former equation gives a value of 428 nm/RIU. The allowed sensing range before interfering with the neighbor resonance peak (equivalent to the free spectral range of the resonator) is 3.45 nm, which is equivalent to about 0.008 RIU change in the test liquid RI. In can be noticed from Equation (6) that the ratio between the length containing the test liquid (the inner diameter of the microtube in our case) to the total cavity length (d_t/d) is best to be 1 (or the closest to 1) for having the highest sensitivity. In our case $d_t/d = 75$ μm/280 μm = 26.8% only. This renders the obtained sensitivity in our case lower than some other Fabry-Pérot cavities even with straight mirrors [15,23,24], but with the use the power drop technique rather than the conventional method of detecting the peak's wavelength shift only, superior sensitivities can be attained as detailed hereafter.

This could be due to the high *Q*-factor that exceeded 2800, which is the highest value reported for an on-chip Fabry-Pérot refractometer.

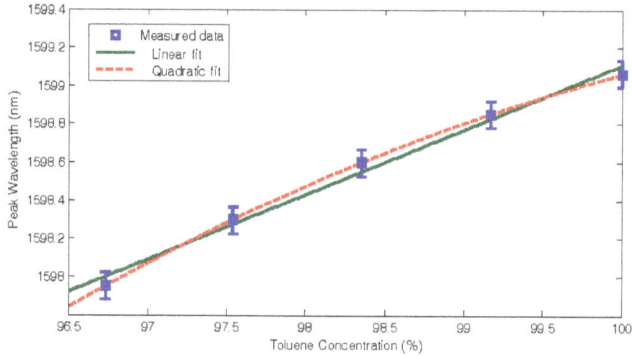

Figure 11. The position of the maxima wavelength *versus* the toluene concentration in the toluene-acetone mixture.

The calculated RI of the unknown mixture obtained by the wavelength shift method is employed to calibrate the sensor operating in the mode of tracing the power drop, to get its sensitivity $\delta P/\delta n_t$. Figure 12 shows the obtained RI values *versus* Toluene concentration, a good agreement between both methods is obtained at $\delta P/\delta n_t$ of approximately 5500 µW/RIU. The range in this case is $-2.73\ \mu\text{W} < \Delta P < -12.12\ \mu\text{W}$, that is equivalent to $0.0005 < \Delta n < 0.0022$. Note that the last point is far from the linear region, and hence it does not fit with the expected RI value.

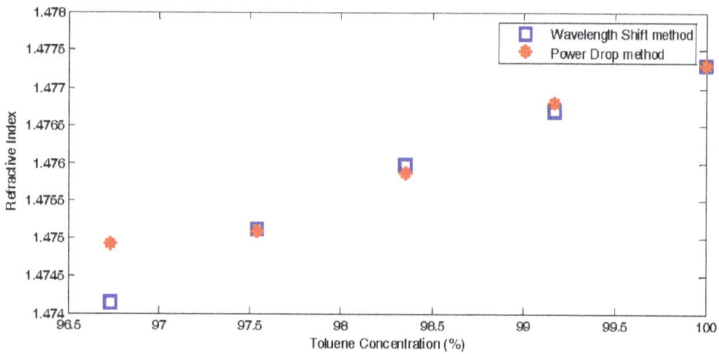

Figure 12. The estimated refractive index *versus* the toluene concentration in the toluene-acetone mixture.

6. Conclusions

Designing a stable Fabry–Pérot cavity employing in-plan cylindrical mirrors and out-of-plan curved surface of a micro-tube has been detailed in this article. By the well confinement of the light inside the cavity attained by the proper design, interference spectrum with high *Q*-factor resonance peaks can be achieved. Such narrow spectral peaks with fast roll-off are useful for sensing applications. On the other hand, such resonator with curved surfaces exhibits higher order resonance modes which may interfere with the principle resonance peak lowering the *Q*-factor as investigated by numerical simulations and observed experimentally. A proper choice of the resonance peak is then critical to achieve high performance sensor. The experimental testing using mixtures of toluene and acetone has been done. By tracing the peak maxima shift in wavelength upon changing the analyte RI, sensitivity up to 428 nm/RIU is achieved along a range of 3.45 nm. High sensitivity up to 5500 µW/RIU can be reached by employing the technique of power tracing at a fixed wavelength, but on a limited range of $0.0005 < \Delta n < 0.0022$. The later method has the advantage of working at a single wavelength and requiring only an optical detector, without the need for sophisticated spectrometry equipment after the device calibration. Noting that, this technique is useful only when absorption difference between the analytes is not significant.

Acknowledgments: The author would like to thank Mme Martine Capo-chichi for providing the absorption data for the liquid mixtures of toluene and acetone using the spectrophotometer in Bâtiment Lavoisier, Université Paris-Est Marne-la-Vallée.

Author Contributions: Noha Gaber performed the stability analysis, numerical simulations, and experiments and wrote most of the paper. Yasser Sabry performed the analytical modeling of Figure 1 and participated in writing the paper. Frédéric Marty fabricated the silicon chips. Tarik Bourouina proposed the architecture of the optofluidic Fabry-Pérot micro-cavity comprising curved surfaces and the use of this cavity as a refractometer and supervised the work.

Conflicts of Interest: The authors declare no conflict of interest.

References

1. Liu, P.Y.; Chin, L.K.; Ser, W.; Chen, H.F.; Hsieh, C.-M.; Lee, C.-H.; Sung, K.-B.; Ayi, T.C.; Yap, P.H.; Liedberg, B.; *et al.* Cell refractive index for cell biology and disease diagnosis: Past, present and future. *Lab Chip* **2016**, *16*, 634–644. [CrossRef] [PubMed]
2. Shelton, D.P. Refractive index measured by laser beam displacement at $\lambda = 1064$ nm for solvents and deuterated solvents. *Appl. Opt.* **2011**, *50*, 4091–4098. [CrossRef] [PubMed]
3. Chaitavon, K.; Sumriddetchkajorn, S.; Nukeaw, J. Highly sensitive refractive index measurement with a sandwiched single-flow-channel crofluidic chip. *RSC Adv.* **2013**, *3*, 6981–6984. [CrossRef]
4. Chao, C.-Y.; Fung, W.; Guo, L.J. Polymer microring resonators for biochemical sensing applications. *IEEE J. Sel. Top. Quantum Electron.* **2006**, *12*, 134–142. [CrossRef]
5. Gaber, N.; Takemura, Y.; Marty, F.; Khalil, D.; Angelescu, D.; Richalot, E.; Bourouina, T. Volume refractometry of liquids using stable optofluidic Fabry-Pérot resonator with curved surfaces. *J. Micro/Nanolithgr. MEMS MOEMS* **2015**, *14*, 045501. [CrossRef]
6. Homola, J.; Yee, S.S.; Gauglitz, G. Surface plasmon resonance sensors: Review. *Sens. Actuators B Chem.* **1999**, *54*, 3–15. [CrossRef]
7. Schmitt, K.; Schirmer, B.; Hoffmann, C.; Brandenburg, A.; Meyrueis, P. Interferometric biosensor based on planar optical waveguide sensor chips for label-free detection of surface bound bioreactions. *Biosens. Bioelectron.* **2007**, *22*, 2591–2597. [CrossRef] [PubMed]
8. Madani, A.; Kleinert, M.; Stolarek, D.; Zimmermann, L.; Ma, L.; Schmidt, O.G. Vertical optical ring resonators fully integrated with nanophotonic waveguides on silicon-on-insulator substrates. *Opt. Lett.* **2015**, *40*, 3826–3829. [CrossRef] [PubMed]
9. Bernardi, A.; Kiravittaya, S.; Rastelli, A.; Songmuang, R.; Thurmer, D.J.; Benyoucef, M.; Schmidt, O.G. On-chip Si/SiOx microtube refractometer. *Appl. Phys. Lett.* **2008**, *93*, 094106. [CrossRef]
10. Böttner, S.; Li, S.; Trommer, J.; Kiravittaya, S.; Schmidt, O.G. Sharp whispering-gallery modes in rolled-up vertical SiO2 microcavities with quality factors exceeding 5000. *Opt. Lett.* **2012**, *37*, 5136–5138.
11. Harazim, S.M.; Quiñones, V.A.B.; Kiravittaya, S.; Sanchez, S.; Schmidt, O.G. Lab-in-a-tube: On-chip integration of glass optofluidic ring resonators for label-free sensing applications. *Lab Chip* **2012**, *12*, 2649–2655. [CrossRef] [PubMed]
12. Zhang, X.; Ren, L.; Wu, X.; Li, H.; Liu, L.; Xu, L. Coupled optofluidic ring laser for ultrahigh sensitive sensing. *Opt. Express* **2011**, *19*, 22242–22247. [CrossRef] [PubMed]
13. Domachuk, P.; Littler, I.; Cronin-Golomb, M.; Eggleton, B. Compact Resonant Integrated Microfluidic Refractometer. *Appl. Phys. Lett.* **2006**, *88*, 093513. [CrossRef]
14. Song, W.; Zhang, X.; Liu, A.; Lim, C.; Yap, P.; Hosseini, H. Refractive index measurement of single living cells using on-chip Fabry-Pérot cavity. *Appl. Phys. Lett.* **2006**, *89*, 203–901. [CrossRef]
15. St-Gelais, R.; Masson, J.; Peter, Y.-A. All-silicon integrated Fabry-Pérot cavity for volume refractive index measurement in microfluidic systems. *Appl. Phys. Lett.* **2009**, *94*, 243905. [CrossRef]

16. Sabry, Y.M.; Saadany, B.; Khalil, D.; Bourouina, T. Silicon micromirrors with three-dimensional curvature enabling lens-less efficient coupling of free-space light. *Light Sci. Appl.* **2013**, *2*, e94. [CrossRef]

17. Sabry, Y.M.; Khalil, D.; Saadany, B.; Bourouina, T. In-plane external fiber Fabry-Pérot cavity comprising silicon micromachined concave mirror. *J. Micro/Nanolithgr. MEMS MOEMS* **2014**, *13*, 011110. [CrossRef]

18. Saleh, B.E.A.; Teich, M.C. *Fundamentals of Photonics*; John Wiley & Sons, Inc.: New York, NY, USA, 1991; pp. 327–330.

19. Yariv, A. *Quantum Electronics*, 3rd ed.; Wiley: New York, NY, USA, 1989; pp. 136–147.

20. Malak, M.; Gaber, N.; Marty, F.; Pavy, N.; Richalot, E.; Bourouina, T. Analysis of Fabry-Pérot optical micro-cavities based on coating-free all-Silicon cylindrical Bragg reflectors. *Opt. Express* **2013**, *21*, 2378–2392. [CrossRef] [PubMed]

21. Mita, Y.; Sugiyama, M.; Kubota, M.; Marty, F.; Bourouina, T.; Shibata, T. Aspect Ratio Dependent Scalloping Attenuation in DRIE and an Application to Low-Loss Fiber-Optical Switches. In Proceedings of the 19th IEEE International Conference on Micro Electro Mechanical Systems (MEMS 2006), Istanbul, Turkey, 22–26 January 2006; pp. 114–117.

22. Kurtz, S.S.; Wikingsson, A.E.; Camin, D.L.; Thompson, A.R. Refractive Index and Density of Acetone-Water Solutions. *J. Chem. Eng. Data* **1965**, *10*, 330–334. [CrossRef]

23. Wei, T.; Han, Y.; Li, Y.; Tsai, H.-L.; Xiao, H. Temperature-insensitive miniaturized fiber inline Fabry-Pérot interferometer for highly sensitive refractive index measurement. *Opt. Express* **2008**, *16*, 5764–5769. [CrossRef] [PubMed]

24. Liu, P.; Huang, H.; Cao, T.; Tang, Z.; Liu, X.; Qi, Z.; Ren, M.; Wu, H. An optofluidics biosensor consisted of high-finesse Fabry-Pérot resonator and micro-fluidic channel. *Appl. Phys. Lett.* **2012**, *100*, 233705. [CrossRef]

micromachines

MDPI

Article

Two-Layer Microstructures Fabricated by One-Step Anisotropic Wet Etching of Si in KOH Solution [†]

Han Lu, Hua Zhang, Mingliang Jin *, Tao He, Guofu Zhou and Lingling Shui *

Institute of Electronic Paper Displays, South China Academy of Advanced Optoelectronics,
South China Normal University, Guangzhou 510006, China; hanlunkq@gmail.com (H.L.);
huazhang717@gmail.com (H.Z.); taohewhh@gmail.com (T.H.); zhougf@scnu.edu.cn (G.Z.)
*   Correspondence: jinml@scnu.edu.cn (M.J.); shuill@m.scnu.edu.cn (L.S.);
    Tel.: +86-20-3931-0068 (M.J.); +86-20-3931-4813 (L.S.); Fax: +86-20-3931-4813 (M.J. & L.S.)
†   This paper is an extended version of our paper published in the 5th International Conference on
    Optofluidics 2015, Taipei, Taiwan, 26–29 July 2015.

Academic Editors: Shih-Kang Fan, Da-Jeng Yao and Yi-Chung Tung
Received: 1 December 2015; Accepted: 18 January 2016; Published: 25 January 2016

Abstract: Anisotropic etching of silicon in potassium hydroxide (KOH) is an important technology in micromachining. The residue deposition from KOH etching of Si is typically regarded as a disadvantage of this technology. In this report, we make use of this residue as a second masking layer to fabricate two-layer complex structures. Square patterns with size in the range of 15–150 μm and gap distance of 5 μm have been designed and tested. The residue masking layer appears when the substrate is over-etched in hydrofluoric acid (HF) solution over a threshold. The two-layer structures of micropyramids surrounded by wall-like structures are obtained according to the two different masking layers of SiO_2 and residue. The residue masking layer is stable and can survive over KOH etching for long time to achieve deep Si etching. The process parameters of etchant concentration, temperature, etching time and pattern size have been investigated. With well-controlled two-layer structures, useful structures could be designed for applications in plasmonic and microfluidic devices in the future.

Keywords: wet etching; potassium hydroxide; Si; pattern

1. Introduction

Anisotropic etching of silicon in alkali metal hydroxides aqueous solutions (e.g., KOH) is an important technology in micromachining [1]. Micro- and nano-structures on substrate have been widely applied in solar cells, superhydrophobic surface and plasmonics [2–6]. Different applications require the features at a different shape and size range.

The formation of micro- and nano-pyramids on the Si surfaces are well known fabricated by this anisotropic etching process [5,6]. KOH has often been selected as the etchant according to its advantages of easy-preparation, non-toxicity, cost-effective and fast-etching [7]. For anisotropic etching of Si, the KOH concentration and etching temperature are key parameters, especially for the structures at nanometer scale. For most micron scale fabrication, the KOH concentration varies from 25–50 wt % and temperature range of 50–85 °C [1,5–16]. The higher etchant concentration and reaction temperature guarantee that the etching products are dissolved fast without hiding the continuous etching. However, if people would like to precisely control the etching process, such as nanostructure by KOH etching, the lower KOH concentration and lower temperature are required [6].

The residue deposition from KOH etching of Si is well known [8], which is typically considered as a disadvantage of this fabrication technology. In this work, we make use of the residue as a second masking layer for fabrication of two-layer microstructures. Square arrays with different pattern size from 15 to 150 μm and gap distance of 5 μm have been designed and tested. Normal micropyramids can be fabricated by well-controlled SiO_2 and Si etched in HF and KOH solutions. When the substrate is over-etched in HF to achieve larger gaps, the second layer wall-like structures appear among the first layer micropyramids. Carefully control the fabrication parameters, the two-layer structure dimensions can be tuned precisely. The residue masking layer is strong and stable, can survive longer than the SiO_2 masking layer for KOH etching, which in the end induces inversed structures.

Two-layer micro/nanostructures are important to achieve step-emulsification microfluidic devices [17,18] and superhydrophobic surfaces [19,20]. With standard microfabrication technology, it typically requires twice lithography and wet-etching to achieve two-layer microstructures. By this method, two-layer microstructures can be fabricated in simple one-step lithography and wet-etching process, which is very useful for fabrication of functional devices for plasmonics and microfluidics applications.

2. Experimental Section

P-type <100> 4″ silicon wafer with 100 nm SiO_2 (Lijing Optoelctronics Co. Ltd, Suzhou, China) was used as the substrate for surface patterning and etching. The Si wafers were ultrasonically cleaned in Deionized (DI) water for 15 min, immersed in Piranha solution for 15 min and then thoroughly rinsed by DI water. Photolithography was done by spin coating (Smart Coater 100, Best Tools, LLC, St Louis, MO, USA) photoresist SUN-120P (Suntific Microelectronic Materials Co. Ltd, Weifang, China) at 3000 rpm for 60 s, exposing using an aligner (URE-200/35, Institute of Optics and Eltcronics, Chendu, China) for 30 s, and developing in 0.4 wt % KOH at 25 °C for 2 min. The wafer with photoresist was then rinsed using DI water and dried using nitrogen gun, and then hard baked on a hot plate (EH20B, Lab Tech, Beijing, China) at 120 °C for 15 min. SiO_2 etching was performed in 10 wt % HF (Guangzhou Chemistry, Guangzhou, China) solution, and anisotropic Si etching was completed in KOH (Zhiyuan Chemistry, Tianjin, China) solution. All chemicals were used as received without further treatment. Desktop scanning electron microscope (SEM) (Phenom G2 Pro, Phenom-World, Eindhoven, The Netherlands) and Field Emission-SEM (FE-SEM) (ZEISS-Ultra55, Carl Zeiss AG, Oberkochen, Germany) were used to visualize micro- and nano-structures and take images. Contact angle was measured using OCS 15pro (Dataphysics, Stuttgart, Germany).

3. Results and Discussion

Pyramids in the range of micrometer to nanometer size can be fabricated by anisotropic etching of silicon surface in KOH etchant solution. The formation of pyramids is neither related to any specific KOH supplier nor to mask or lithography problems [1]. Usually, the KOH etched pyramids are of exact geometric shape. Two types of pyramids can be observed: rectangular base or octagonal base, depending on the experimental conditions [21].

In general, pyramids are obtained according to the anisotropic etching of <100> and <110> plane in KOH solution. In this work, we have designed the square patterns in the range of 15 to 150 μm with gap distance of 5 μm. By controlling the etching time in HF solution, the opening of SiO_2 (first masking layer for KOH etching) varies with etching time. We have obtained simple micropyramids structure by strictly control the etching time in HF solution. However, prolonged etching time in HF solution expand the opening size of the SiO_2 on Si substrate, which causes fast reaction with large amount of products which will deposit on Si surface serving as a second masking layer to produce a second layer of wall-like structures among the first layer micropyramids.

3.1. Two Types of Microstructures Obtained in One-Step Wet Etching

In our experiments, we found that with the variation of pattern size and etching conditions different types of patterns have been obtained. Figure 1a,c represents the schematic cross-sectional view of the fabrication process of one-layer micropyramids and two-layer with first layer micropyramids surrounded by second layer wall-like structures, respectively. Figure 1b,d shows the SEM images of the fabricated structures.

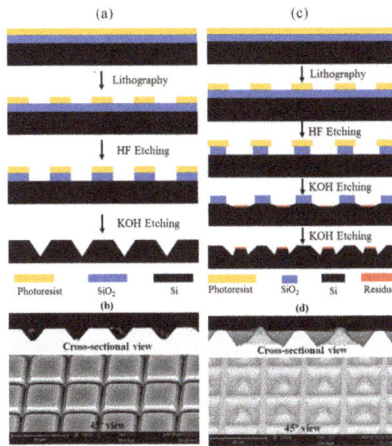

Figure 1. Schematic drawing of the fabrication process without (**a**) or with (**c**) the residue as second masking layer. (**b**) SEM images of cross-sectional view (top) and 45° view (bottom) of the fabricated one-layer micropyramids. (**d**) SEM images of cross-sectional view (top) and 45° view (bottom) of the fabricated two-layer micropyramids surrounded by walls.

The simple micropyramids are obtained by anisotropic etching silicon via the 5 μm opening in KOH solution, micropyramids connected with each other via the valleys with the same shape and size, as shown in Figure 1b. By carefully analyzing SEM image in Figure 1d (top), we can clearly see that the height and width of the neighboring micropyramids are different; however, all even micropyramids show the same shape and size and all odd micropyramids show the same shape and size. The bottom image in Figure 1d clearly show that two types of microstructures were obtained in our experiments. The first layer of micropyramids were obtained by anisotropic etching silicon substrate via the opened Si by HF etching of SiO$_2$, and the second layer of wall-like structures surrounding the micropyramids were obtained according to the second masking layer from the reaction products deposition. The second residue masking layer is mainly Si(OH)x from the reaction of KOH and Si, which could be easily removed if the sample was dipped in HF solution again. Figure 2 shows the high resolution SEM image of the substance deposited on the wall-like structure surface.

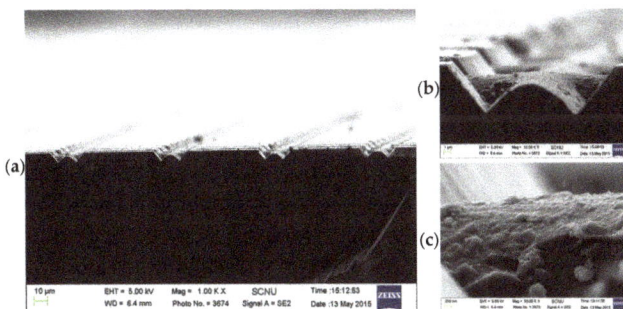

Figure 2. SEM images of a two-layer structures (**a,b**), and the residue substance on the wall-like structure surface (**c**). The pattern size was 75 μm × 75 μm with 5 μm gap distance. The substrate was etched in 10 wt % HF for 1.5 min, and then in 10 wt % KOH solution for 30 min at 70 °C.

We have designed and tested the square patterns with side length of 15, 30, 50, 65, 75, 100 and 150 µm, and all gap distance of 5 µm. Both simple one-layer micropyramids and complex two-layer structures of micropyramids surrounded by the wall structures have been observed for all these designed structures. In general, the one-layer micropyramids appear when the etching time in HF solution (t_{HF}) is ⩽ 1.0 min. As soon as t_{HF} ⩾ 1.5 min, the two-layer structures started to appear.

3.2. HF Etching Time Effect on the Two-Layer Structures

As discussed in the previous section, the two-layer structures appear according to the over etching in HF solution. We have tested the HF etching time (t_{HF}) effect on the first layer micropyramids and second layer wall-like structures using the sample of 50 µm × 50 µm square patterns with 5 µm gap distance, as shown in Figure 3. Figure 3a is the SEM images of the fabricated structure when the samples were immersed in HF for different time. Figure 3b,c shows the variation of the two-layer structure width and height with t_{HF}.

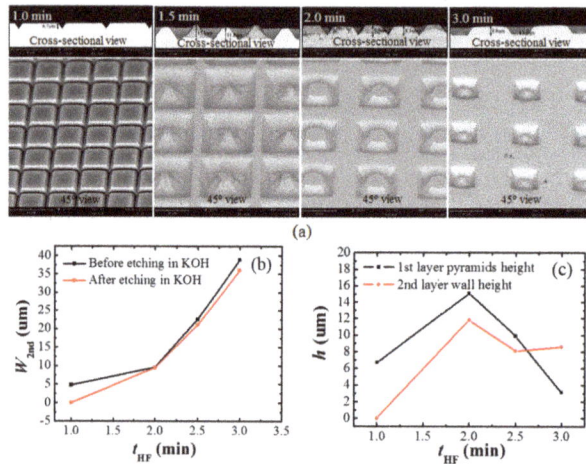

Figure 3. (**a**) SEM images of the fabricated structures at different t_{HF} (top: cross-sectional view, bottom: 45° view). (**b**) The second layer wall width (W_{2nd}) varies with t_{HF}. (**c**) Height (h) of the first layer micropyramids and the second layer walls changes with t_{HF}. The sample patterns are 50 µm × 50 µm square with gap distance of 5 µm. The samples were etched in HF solution for 1.0, 1.5, 2.0 and 2.5 min. All sample substrates were etched in 10 wt % KOH solution for 30 min at 70 °C.

The simple one-layer micropyramids were obtained when the sample was etched in HF for 1.0 min. As t_{HF} increased to 1.5, 2.0 and 3.0 min, obvious two-layer structures were observed. The height of the micropyramids (first layer) and wall-like structures (second layer) is 15.1 and 11.8 µm when t_{HF} = 1.5 min, 9.9 and 8.1 µm when t_{HF} = 2.0 min, 3.1 and 8.6 µm when t_{HF} = 3.0 min. Therefore, the height difference between first layer and second layer is 3.3, 1.8 and 5.5 µm for t_{HF} of 1.5, 2.0 and 3.0 min, respectively. The maximum second layer wall-like structures were obtained at t_{HF} = 1.5 min. A special structure was obtained at t_{HF} = 3.0 min, in which the micropyramids were smaller than the wall structures. As seen from the details of the SEM images, the SiO_2 layer disappeared during the KOH etching, leaving its covered Si completely open for KOH etching which caused the micropyramids to shrink. However, the wall-like structures were stable over all etching processes, showing strong and wide wall structures. Therefore, we can conclude that the two-layer complex structures are produced according to the two masking layers: the first SiO_2 masking layer and the second residue masking layer from the quick accumulation of reaction products of KOH and Si in the open area, as demonstrated in Figure 1c.

3.3. KOH Concentration and Etching Time Effect on Two-Layer Structures

The effect of KOH concentration (C_{KOH}) and etching time in KOH solution (t_{KOH}) on the second layer wall structures have also been investigated, as shown in Figure 4. As the KOH concentration increases, the wall width does not change significantly, as shown in Figure 4a. The wall width slightly decreases with the etching time in KOH, as shown in Figure 4b.

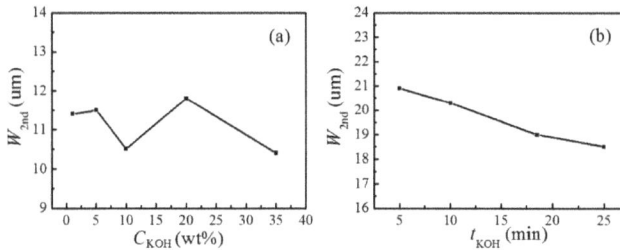

Figure 4. (a) Width of the second layer walls (W_{2nd}) varies with KOH concentration(C_{KOH}). (b) Width of the second layer walls (W_{2nd}) changes with etching time in KOH solution (t_{KOH}). The sample patterns are 50 μm × 50 μm square with gap distance of 5 μm. Each data point was obtained by averaging the values from three samples. All samples were etched in HF solution for 1.5 min. The etching time and temperature was 30 min and 70 °C for (a). The KOH concentration was 10 wt % KOH and etching temperature was 70 °C for (b).

3.4. Shapes of Micropyramids

Usually the KOH etched pyramids are of exact geometric shape. Two types of pyramids can normally be observed: rectangular base or octagonal base, depending on the experimental conditions [21]. In our experiments, we have also observed the rectangular and octagonal shapes for the first layer micropyramids directly etched via the opening of the SiO_2 layer. The samples with the same pattern size (50 μm × 50 μm square with 5 μm gap distance) was selected to investigate the shape of the micropyramids. The samples were all etched in 10 wt % HF solution for 2.0 min, and then moved to KOH solutions to etch for 10 min at 70 °C. As shown in Figure 5, two-layer structures are obtained for all samples; however the shape of the micropyramid tips changes with process conditions. The shape is square, octagonal, mixed square and octagonal, square with round-corners and round at the KOH concentration of 1.0, 5.0, 10, 20 and 35 wt %, respectively. This is according to the plane selectivity and etching speed, which is similar to simple micropyramid structures without the second layer structures [9].

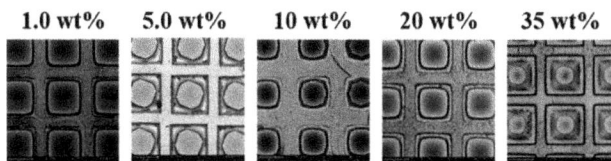

Figure 5. First layer micropyramids shape varies with KOH concentration. All samples were 50 μm × 50 μm square patterns with gap of 5 μm. The samples were first etched in 10 wt % HF solution for 2.0 min. The KOH etching time was 10 min and temperature was 70 °C.

4. Conclusions

In this work, we have designed micropatterns for fabrication of micropyramids on silicon substrate with SiO_2 as masking layer. Simple one-layer micropyramids have been obtained with controlled etching in HF solution to open the etching access holes at around 5 μm. Two-layer complex structures of first layer micropyramids surrounded by second layer wall-like structures were obtained by over-etching the substrate in HF solution. With the square pattern size in the range of 15 to 150 μm with 5 μm gap distance, reproducible two-layer structures have been obtained in a wide range of fabrication conditions. As an example, 50 μm × 50 μm patterns with 5 μm gap, the two layer micropyramids appeared when HF etching time was more than 1.5 min. With increasing the etching time in HF, the second layer wall-like structure width and height increases, however the first layer micropyramids height and width decreases. KOH concentrate does not affect the wall-like structure size significantly; however, it does affect micropyramids shapes due to the plane selectivity and etching speed. The residue masking layer is strong and stable, which can withstand long time KOH etching. For a long time HF etching with less SiO_2 left, the second layer mask can induce an inversed structures after long time etching in KOH solution. Making using of the residue as a second masking layer for controllable microstructure fabrication on Si substrate is a simple and useful method which will be used for fabrication of various structures applied in plasmonics, microfluidics and superhydrophobic surfaces in the future.

Acknowledgments: We appreciate the financial support from the National Nature Science Foundation of China (NSFC No. 61574065 and 21303060). This work was also supported by Program for Changjiang Scholars and Innovative Research Team in University (IRT13064), International Cooperation Base of Infrared Reflection Liquid Crystal Polymers and Device (2015B050501010), Guangdong Innovative Research Team Program (2011D039), and Guangdong Talent Program (201101D0104904202).

Author Contributions: Mingliang Jin and Lingling Shui designed and conducted this project. Han Lu performed the experiments together with Tao He and Hua Zhang. The data summary and writing of the article was mainly done by Han Lu, Mingliang Jin and Lingling Shui. Guofu Zhou gave suggestions on the project management and conducted helpful discussion on the experimental results and manuscript writing.

Conflicts of Interest: The authors declare no conflict of interest.

References

1. Schroder, H.; Obermeier, E.; Steckenborn, A. Micropyramidal hillocks on KOH etched {100} silicon surfaces: Formation, prevention and removal. *J. Micromech. Microeng.* **1999**, *9*, 139–145.
2. Juvonen, T.; Harkonen, J.; Kuivalainen, R. High efficiency single crystalline silicon solar cells. *Phys. Scr.* **2002**, *T101*, 96–98. [CrossRef]
3. Pal, P.; Sato, K.; Shikida, M.; Gosálvez, M.A. Study of corner compensating structures and fabrication of various shapes of MEMS structures in pure and surfactant added TMAH. *Sens. Actuators A Phys.* **2009**, *154*, 192–203. [CrossRef]
4. Berim, G.O.; Ruckenstein, E. Contact angle of a nanodrop on a nanorough solid surface. *Nanoscale* **2015**, *7*, 3088–3099. [CrossRef] [PubMed]
5. Kumar, M.D.; Kim, H.; Kim, J. Periodically patterned Si pyramids for realizing high efficient solar cells by wet etching process. *Sol. Energy* **2015**, *117*, 180–186. [CrossRef]
6. Ngoc, L.L.T.; Jin, M.; Wiedemair, J.; van den Berg, A.; Carlen, E.T. Large area metal nanowire arrays with tunable sub-20 nm nanogaps. *ACS Nano* **2013**, *7*, 5223–5234. [CrossRef] [PubMed]
7. Dutta, S.; Imran, M.; Kumar, P.; Pal, R.; Datta, P.; Chatterjee, R. Comparison of etch characteristics of KOH, TMAH and EDP for bulk micromachining of silicon (110). *Microsys. Technol.* **2011**, *17*, 1621–1628. [CrossRef]
8. Seidel, H. The mechanism of anisotropic, electrochemical silicon etching in alkaline solutions. In Technical Digest of the 4th IEEE Solid-State Sensor and Actuator Workshop, Hilton Head Island, SC, USA, 4–7 June 1990; pp. 86–91.
9. Seidel, H.; Csepregi, L.; Heuberger, A.; Baumgärtel, A. Anisotropic etching of crystalline silicon in alkaline solutions. I. Orientation dependence and behavior of passivation layers. *J. Electrochem. Soc.* **1990**, *137*, 3612–3626.
10. Williams, K.R.; Muller, R.S. Etch rates for micromachining processing. *J. Microelectromech. Syst.* **1996**, *5*, 256–269. [CrossRef]

11. Tokoro, K.; Uchikawa, D.; Shikida, M.; Sato, K. Anisotropic etching properties of silicon in KOH and TMAH solutions. In Proceedings of the 1998 International Symposium on Micromechatronics and Human Science (MHS '98), Nagoya, Janpan, 25–28 November 1998; pp. 65–70.
12. Chen, P.-H.; Peng, H.-Y.; Hsieh, C.-M.; Chyu, M.K. The characteristic behavior of TMAH water solution for anisotropic etching on both silicon substrate and SiO$_2$ layer. *Sens. Actuators A Phys.* **2001**, *93*, 132–137. [CrossRef]
13. Yonghao, X.; Lingbo, Z.; Dennis, W.H.; Wong, C.P. Hierarchical silicon etched structures for controlled hydrophobicity/superhydrophobicity. *Nano Lett.* **2007**, *7*, 3388–3393.
14. Muñoz, D.; Carreras, P.; Escarré, J.; Ibarz, D.; de NicoLás, S.M.; Voz, C.; Asensi, J.M.; Bertumeu, J. Optimization of KOH etching process to obtain textured substrates suitable for heterojunction solar cells fabricated by HWCVD. *Thin Solid Films* **2009**, *517*, 3578–3580. [CrossRef]
15. Fan, Y.; Han, P.; Liang, P.; Xing, Y.; Ye, Z.; Hu, H. Differences in etching characteristics of TMAH and KOH on preparing inverted pyramids for silicon solar cells. *Appl. Surf. Sci.* **2013**, *264*, 761–766. [CrossRef]
16. Branham, M.S.; Hsu, W.-C.; Yerci, S.; Loomis, J.; Boriskina, S.V.; Hoard, B.R.; Han, S.E.; Chen, G. 15.7% efficient 10-μm-thick crystalline silicon solar cells using periodic nanostructures. *Adv. Mater.* **2015**, *27*, 2182–2188. [CrossRef] [PubMed]
17. Shui, L.; Mugele, F.; van den Berg, A.; Eijkel, J.C.T. Geometry-controlled droplet generation in head-on microfluidic devices. *Appl. Phys. Lett.* **2008**, *93*, 153113. [CrossRef]
18. Arayanarakool, R.; Shui, L.; Kengen, S.W.M.; van den Berg, A.; Eijkel, J.C.T. Single-enzyme analysis in a droplet-based micro- and nanofluidic system. *Lab Chip* **2013**, *13*, 1955–1962. [CrossRef] [PubMed]
19. Wu, X.F.; Shi, G.Q. Fabrication of a lotus-like micro-nanoscale binary structured surface and wettability modulation from superhydrophilic to superhydrophobic. *Nanotechnology* **2005**, *16*, 2056–2060. [CrossRef] [PubMed]
20. Li, X.G.; Shen, J. A facile two-step dipping process based on two silica systems for a superhydrophobic surface. *Chem. Commun.* **2011**, *47*, 10761–10763. [CrossRef] [PubMed]
21. Zubel, I.; Barycka, I. Silicon anisotropic etching in alkaline solutions I. The geometric description of figures developed under etching Si (100) in various solutions. *Sens. Actuators A Phys.* **1998**, *70*, 250–259. [CrossRef]

micromachines

MDPI

Article

Fluid Flow Shear Stress Stimulation on a Multiplex Microfluidic Device for Rat Bone Marrow Stromal Cell Differentiation Enhancement [†]

Chia-Wen Tsao [1,]*, Yu-Che Cheng [2,3,4] and Jhih-Hao Cheng [1]

[1]   Department of Mechanical Engineering, National Central University, 32001 Taoyuan, Taiwan;
      uf2lab@gmail.com
[2]   Proteomics Laboratory, Cathay General Hospital, 22174 New Taipei City, Taiwan; yccheng@cgh.org.tw
[3]   Institute of Biomedical Engineering, National Central University, 32001 Taoyuan, Taiwan
[4]   School of Medicine, Fu Jen Catholic University, 24205 New Taipei City, Taiwan
*   Correspondence: cwtsao@ncu.edu.tw; Tel.: +886-3-426-7343; Fax: +886-3-425-4501
†   This paper is an extended version of our paper presented in the 5th International Conference on
      Optofluidics 2015, Taipei, Taiwan, 26–29 July 2015.

Academic Editors: Shih-Kang Fan, Da-Jeng Yao and Yi-Chung Tung
Received: 28 October 2015; Accepted: 7 December 2015; Published: 11 December 2015

Abstract: Microfluidic devices provide low sample consumption, high throughput, high integration, and good environment controllability advantages. An alternative to conventional bioreactors, microfluidic devices are a simple and effective platform for stem cell investigations. In this study, we describe the design of a microfluidic device as a chemical and mechanical shear stress bioreactor to stimulate rat bone marrow stromal cells (rBMSCs) into neuronal cells. 1-methyl-3-isobutylxanthine (IBMX) was used as a chemical reagent to induce rBMSCs differentiation into neurons. Furthermore, the shear stress applied to rBMSCs was generated by laminar microflow in the microchannel. Four parallel microfluidic chambers were designed to provide a multiplex culture platform, and both the microfluidic chamber-to-chamber, as well as microfluidic device-to-device, culture stability were evaluated. Our research shows that rBMSCs were uniformly cultured in the microfluidic device and differentiated into neuronal cells with IBMX induction. A three-fold increase in the neuronal cell differentiation ratio was noted when rBMSCs were subjected to both IBMX and fluid flow shear stress stimulation. Here, we propose a microfluidic device which is capable of providing chemical and physical stimulation, and could accelerate neuronal cell differentiation from bone marrow stromal cells.

Keywords: microfluidics; rat bone marrow stromal cell; stem cell stimulation; fluid flow shear stress; neuronal cell differentiation

1. Introduction

Microfluidic devices, an integrated system incorporate various function such as pumping, mixing, sample separation, sample concentration, and culturing in a single chip for chemical or biological analysis. Micro total analysis system (µTAS) or lab-on-a-chip (LOC) are sometimes also referred to as microfluidic technologies. In recent decades, microfluidics has become a widely used platform for cell investigations. Compared to conventional cell culture techniques in a Petri dish, microfluidics offers low sample/reagent consumption, which offers high integration, high automation, and a real-time monitoring cell culture approach [1]. Due to these advantages, various unique microfluidic cell investigations have been demonstrated. For example, Emneus *et al.* developed a long-term, real-time microfluidic device to monitor human cells (HFF11) [2]. Ramsey's group integrated cell separation and mass spectrometry in a chip for cell analysis [3]. Single-cell sorting and analysis can be achieved based on droplet microfluidics [4]. Polydimethylsiloxane (PDMS) soft elastomer is the most commonly used material in cell culture microfluidic devices because of its good biocompatibility, gas permeability, and optical transmissivity [5,6]. Since PDMS-based microfluidic devices are mainly fabricated by a soft lithography process, the cell culture microchannel geometry can be precisely controlled and tailored to various shapes for cell culture investigations [7]. With further integration of PDMS microfluidic valves, high throughput screening can be achieved in such microfluidic devices [8]. Besides, conventional Petri dish is a two-dimensional environment. Microfluidics provides a three-dimensional microenvironment that provide more *in vivo* conditions for cell investigations [9]. Recently, microfluidics devices have moved to organs-on-chips [10,11] applications, which use microfluidic techniques to mimic organs in biochemical microenvironment for *in vitro* disease models.

Of the research conducted in cell investigations, stem cells have the ability of self-renewal and can differentiate into specific cell types through environmental stimuli. Due to their potential to regenerate and repair damaged tissue, stem cells have been applied to various promising applications, such as tissue engineering or cell-based therapies. Chemical and physical factors, which influence stem cell differentiation have been extensively studied and are known to have significant effects on cell differentiation [12]. Recently, in addition to chemical signals, mechanical forces are also found to affect stem cell differentiation. Mechanical forces, such as hydrostatic pressure [13–15], shear stress [16,17], and tensile strain [18,19] have critical roles in deciding stem cell fate. While conventional methods studying mechanical force effects normally require complex, custom-made bioreactor designs, microfluidic devices offer a simple, low-cost alternative choice for stem cell investigations [20,21]. Microfluidic devices provide good environmental regulatory ability to mimic *in vivo* microenvironment, which is ideal for stem cell investigations. Since microfluidic devices are a dynamic profusion-based environment, liquid phase chemical reagents can simply be replaced or injected into the microchannel for chemical stimuli research. Research in cell chemotaxis can be easily performed while further integrating a microfluidic gradient generator [22]. In mechanical stress stimulation experiments, a conventional bioreactor uses a compressive loading cylinder [23], rotating cone [24,25], or parallel plate chamber [26] to generate hydrostatic pressure or shear stress on cells. With a microfluidic device, on the other hand, physical shear stress can be simply created by fluid flow induction from laminar microflow behavior [27], and hydrostatic pressure can be created by a vortex in the PDMS chamber [28]. These examples show that the microfluidic device is a simple and effective approach to study the mechanical stress effects on stem cells. Therefore, in this study, we used a microfluidic device as a chemical and mechanical shear stress bioreactor to stimulate rat bone marrow stromal cells (rBMSCs). The effects of chemical and mechanical shear stress on the rBMSCs were reported in this investigation. Four parallel microfluidic rBMSCs cultures and differentiation chamber designs were used in this study to highlight potential of high-throughput screening.

2. Experimental Section

2.1. Materials and Reagents

Materials and reagents for microfluidic device fabrication are listed as below. A 10-cm diameter P-type (100) single-side polished silicon wafer with a resistivity of 1–100 $\Omega \cdot$ cm was purchased from Summit-Tech Resource Corp. (Hsinchu, Taiwan). SU-8 3050 kit including SU-8 photoresist and SU-8 developer was purchased from MicroChem Corp. (Newton, MA, USA). PDMS elastomer kit including PDMS base and curing agent (Sylgard 184 silicone elastomer kit) was purchased from Dow Corning Corp. (Midland, MI, USA). Stainless steel surgical needles (SC20/15) were purchased from Instech Laboratories Inc. (Plymouth Meeting, PA, USA). Medical grade plastic tubing (Tygon S-50-HL) were purchased from Saint-Gobain Performance Plastics (Akron, OH, USA). 75 mm × 25 mm × 1 mm microscope glass slides were purchased from Doger Instruments Co. Ltd (Taipei, Taiwan). Acetone (ACE) and isopropyl alcohol (IPA) were purchased from J.T. Baker (Phillipsburg, NJ, USA).

Materials and reagents for rBMSCs culturing and differentiation experiments are listed as below. Culture Petri dishes (Nunc) were purchased from Thermo Fisher Scientific (Waltham, MA, USA). Primary antibodies against the human neuron-specific enolase (NSE, 1:25) and 1-methyl-3-isobutylxanthine (IBMX) were purchased from Sigma-Aldrich (St. Louis, MO, USA). Avidin-biotin conjugate of horseradish peroxidase and Vector VIP substrate kit were purchased form Vector Laboratories (Burlingame, CA, USA). Dulbecco's modified Eagle medium (DMEM) and Penicillin/Streptomycin (P/S) were purchased from Gibco/Life Technologies (Carlsbad, CA, USA). Fetal Bovine Serum (FBS) were purchased from HyClone/GE Healthcare (Novato, CA, USA).

2.2. Rat Bone Marrow Stromal Cell Preparation

rBMSCs were harvested from eight-week-old Sprague–Dawley rats. Cell cultures were maintained at 37 °C with 5% CO_2 in a 10 cm diameter culture dish with culture medium which consisted of DMEM with 10% FBS, 100 U/mL penicillin, and 100 g/mL streptomycin. rBMSCs used in this study maintained in 20–25 passages for strong proliferation potential.

2.3. 4′,6-Diamidino-2-Phenylindole (DAPI) and Immunocytochemistry (ICC) Staining Procedures

4′,6-Diamidino-2-Phenylindole (DAPI) staining was performed on the rBMSCs to ensure accurate cell number calculations for the rBMSCs cell growth experiments. The DAPI staining process is performed directly on the microfluidic bioreactor and the detailed staining procedure is illustrated below. A PBS rinse was performed between each chemical injection step. The following on-chip staining was operated at room temperature and injected with 1.5 μL/min unless described elsewhere. First, we injected 4% paraformaldehyde for 10 min to fix the cells followed by the PBS rinse. Then, we injected 0.1% Triton-X 100 for 4 min to permit cell permeabilization. Finally, diluted DAPI solution (DAPI: PBS, 1:800, v/v) was injected into the microchannel to stain the rBMSCs cell nucleus.

Similar to DAPI staining, the immunocytochemistry (ICC) staining process was performed in this study to confirm the neuronal cell type and ensure accurate neuronal cell calculation. The ICC staining process was also performed directly on the microfluidic bioreactor and the staining procedure was illustrated as follows. The rBMSCs were first treated with paraformaldehyde and Triton-X 100 to fix and permeabilize the cells, which is identical to the DAPI staining process as described above. Then, 3% H_2O_2 solution was injected into the microchannel and allowed to rest for 30 min before 1.5% normal horse serum was injected into the microchannel. This was allowed to rest for 30 min and was followed by a PBS rinse. Next, rBMSCs were treated with primary antibody against the human antigen neuron-specific enolase and incubated at 37 °C for 1 h. After primary antibody treatment, the cells were then stained with biotinylated anti-rabbit antibody followed by an avidin-biotin conjugated horseradish peroxidase for 30 min. Finally, Vector VIP substrate kit was introduced to the microfluidic chamber as a visualization reagent to reveal the resulting peroxidase activity.

2.4. Microfluidic Device Fabrication

Fabrication of the microfluidic device is based on a standard PDMS soft lithography process which requires micromold fabrication and PDMS replication process as shown in Figure 1a–c and Figure 1d–f, respectively. In micromold fabrication, the bare silicon wafer was first cleaned with acetone (ACE), isopropyl alcohol (IPA), and deionized (D.I.) water, followed by baking at 130 °C for 15 min on a hot plate (Super-Nuova, Barnstead Thermolyne, Waltham, MA, USA) to remove the moisture on the silicon surface for enhanced SU-8 attachment. Then, spin coat (SPC-703, Yi YANG Co., Taoyuan, Taiwan) SU-8 photoresist was added to the silicon substrate (Figure 1a). The spin coating speed was set at 500 rpm for 30 s and 1000 rpm for 40 s. After spin coating, the SU-8 layer was subject to UV exposure for 90 s (AGL100 UV Light source, M & R Nano Technology Co., Taoyuan, Taiwan) via a patterned transparent film mask (Ching Acme Enterprises Corp., Taipei, Taiwan) to generate microchannel patterns on the SU-8 layer (Figure 1b). Immersed, UV-exposed SU-8 substrate in SU-8 developer for 3–5 min to create a SU-8 micromold for the subsequent PDMS replication process (Figure 1c). After

SU-8 micromold fabrication, PDMS casting was performed to create the microchannel. The PDMS replication process started with pouring a 10:1 volume ratio of Sylgard 184 base and curing agent mixture over the SU-8 micromold before degassing in a vacuum oven to remove the air bubbles inside the PDMS layer. Then, the PDMS layer was cured at 70 °C for 4 h (Figure 1d), removed from the SU-8 micromold and O_2 plasma bond PDMS with a glass slide (Duffy *et al.* 1998) in O_2 plasma cleaner (PDC-32G, Harrick Plasma, Ithaca, NY, USA) for 150 s (Figure 1e). Punch fluid inlet and outlet holes was conducted using a biopsy punch (Harris Uni-Core, Sigma-Aldrich, St. Louis, MO, USA) and inserted stainless steel surgical needles as microfluidic bioreactors for injection and outlet connectors (Figure 1f). Medical grade plastic tubing was used to connect the needle inject connector to the syringe pump or waste beaker.

All microfluidic channels, plastic tubes, surgical needles, and injection syringes were fully sterilized to prevent cell contamination during experiments. The plastic tube, surgical needles, and injection syringe were cleaned by immersion in 75% alcohol solution followed by UV sterilization at 30 min. After UV sterilization, the microfluidic bioreactor was connected to the syringe pump, and Dulbecco's Phosphate Buffered Saline (DPBS) was flushed into the microchannel for 90 min to clean the microchannel walls.

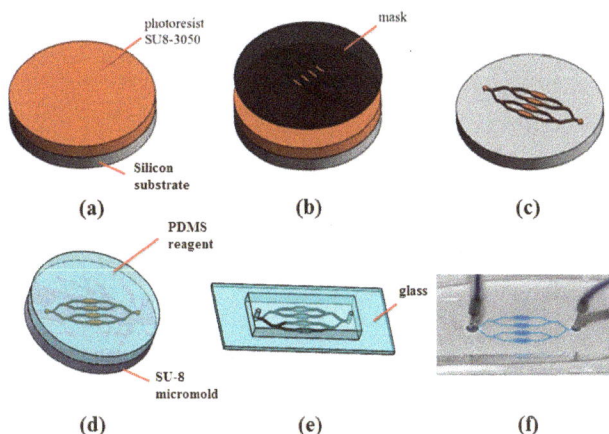

Figure 1. Schematic illustration of SU-8 micromold fabrication procedure: (**a**) silicon wafer cleaning and spin coat SU-8; (**b**) UV exposure; (**c**) SU-8 development and PDMS replication process; (**d**) casting the PDMS on SU-8 micromold; (**e**) O_2 plasma bonding PDMS layer with glass substrate; and (**f**) photography of PDMS microchannel with blue color dye injection.

3. Results and Discussion

3.1. rBMSCs Culture and Stability in the Four Parallel Microfluidic Chambers

The rBMSCs were cultured and differentiated into neuronal cells in a microfluidic device. The microfluidic device design layout and experiment setup are shown in Figure 2. The microchannel consists of a 1 mm diameter inlet and outlet port for surgical stainless needle insertion. The microfluidic device injection inlet was connected to the syringe pump and the outlet was connected to the waste beaker through a plastic surgical tube, respectively. The microchannel is 200 μm in width and 100 μm in height. Two Y-shaped microchannel splitters (first splitter: 44°, second splitter: 25°) split the flow equally into the four microfluidic chambers (dimensions: 4 mm length, 1 mm width, and 0.1 mm height), where rBMSCs culture and differentiation took place. The microfluidic device system was placed inside the CO_2 incubator (SCA-165DRS, ASTEC, Fukuoka, Japan) and maintained at 37 °C and 5% CO_2 concentration.

The rBMSCs culture started with a dynamic injection in the inlet port of ~10 µL of 10^6 cells/mL rBMSCs suspension at 2 µL/min to seed the rBMSCs in the four microfluidic chambers. The seeding procedure was operated under inverted microscope to check the cells are fully injected into each chambers. After rBMSCs seeding, the cells were left in the CO_2 incubator without motion for 24 h under stable conditions, allowing the rBMSCs to fully adhere to the microfluidic chamber. As shown in Figure 3a, around 1 µm diameter rBMSCs suspension can clearly be found and were uniformly distributed inside the microfluidic chamber right after the rBMSCs suspension injection step. As shown in Figure 3b, spindle-like rBMSCs morphology was uniformly distributed inside the microfluidic chamber, which indicates that the rBMSCs were tightly adherent to the microfluidic device substrate.

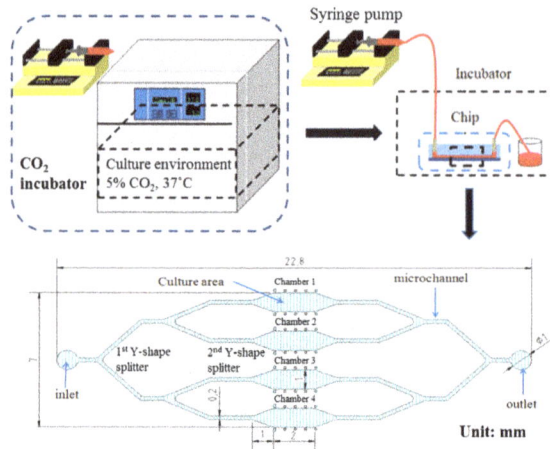

Figure 2. Schematic illustration of rBMSCs fluid flow stimulation microchannel design and experimental setup.

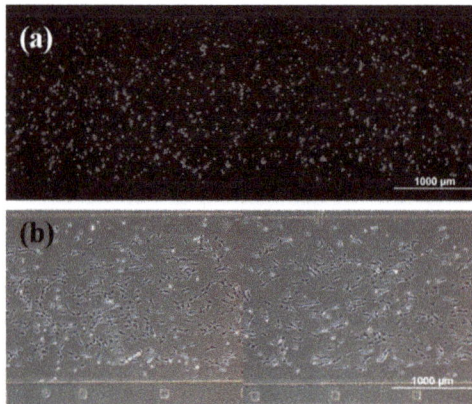

Figure 3. Microscopy images of rBMSCs seeding after (**a**) cell suspension injection and (**b**) 24 h static seeding.

In the high-throughput microfluidic device, cell culture stability is important to ensure good cell culture uniformity, as well as controlled chemical and physical stimulation conditions among all microfluidic chambers. In theory, the hydraulic resistance of four parallel microfluidic channels is identical and delivers equal amounts of rBMSCs and fluid flow shear stress to each microfluidic chamber. However, because of the microfabrication variations, fluid flow as well as the culture conditions for the four parallel microfluidic chamber may vary. Therefore, the microfluidic chamber-to-chamber and device-to-device culture stability were evaluated. Figure 4 shows the bright field microscopic and DAPI staining images of rBMSCs inside the four microfluidic chambers. The rBMSCs were uniformly distributed and cultured inside each individual microfluidic chamber, but the cell number among individual microfluidic chambers was found to be slightly different.

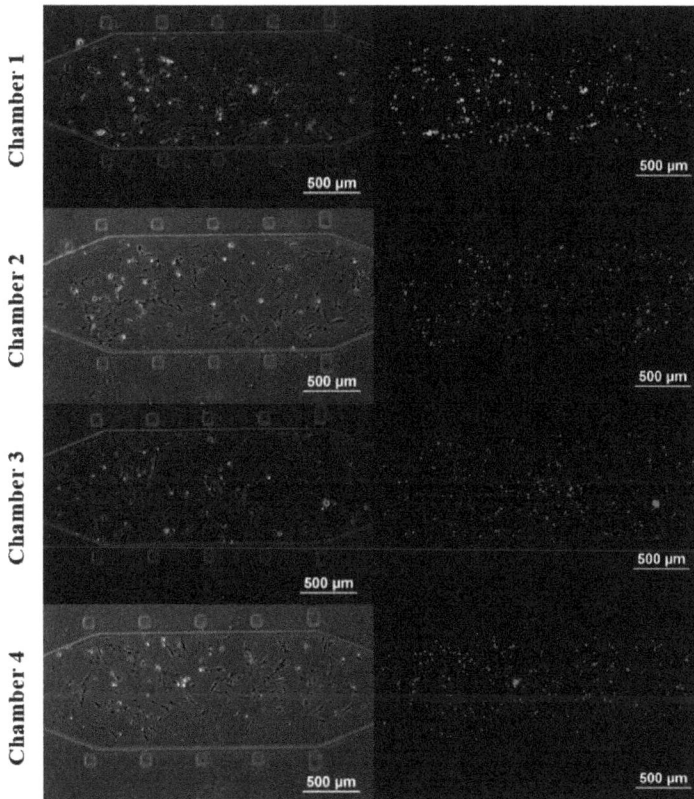

Figure 4. Microscopy (**left**) and DAPI-stained (**right**) images of rBMSCs culturing conditions in the microfluidic chambers 1–4.

To quantify the cell number inside the microfluidic chamber, the DAPI staining images taken from the fluorescence microscope (ECLIPSE Ti-U, Nikon, Tokyo, Japan) were further analyzed by the image analysis software ImageJ to calculate the cell number inside the microfluidic chambers. Figure 5 shows the rBMSCs number from microfluidic chambers 1–4. Four individual microfluidic devices were fabricated to evaluate the microfluidic chamber-to-chamber as well as microfluidic device-to-device variations. In microfluidic Device 1, the DAPI stained rBMSCs inside the microfluidic chamber numbered 138 (Chamber 1), 178 (Chamber 2), 167 (Chamber 3), and 237 (Chamber 4), with an average value of 181 ± 43 cells. Similarly, the rBMSCs inside the microfluidic chambers numbered 156/254/177 (Chamber 1), 140/224/237 (Chamber 2), 175/234/171 (Chamber 3), and 227/207/202 (Chamber 4), with an average value of 175 ± 38/230 ± 20/197 ± 30 cells for microfluidic Device 2/Device 3/Device 4.

From the DAPI stained cell number data, the microfluidic chamber-to-chamber relative standard deviation percentage (RSD%) was measured as 23.5%, 21.7%, 8.5%, and 15.2% for Devices 1–4, respectively. The microfluidic chip-to-chip RSD% value was measured at 28.2%, 22.8%, 17.0%, and 7.9% for Chambers 1–4, respectively. Compared with multiplex fluid flow delivery microfluidic device with 34.7% volume variation between microfluidic channels [29]. The fluid-flow driven microfluidic device shows good stability with an overall average RSD% of 17.3% between microfluidic chambers and an overall average RSD% of 18.9% between microfluidic device events.

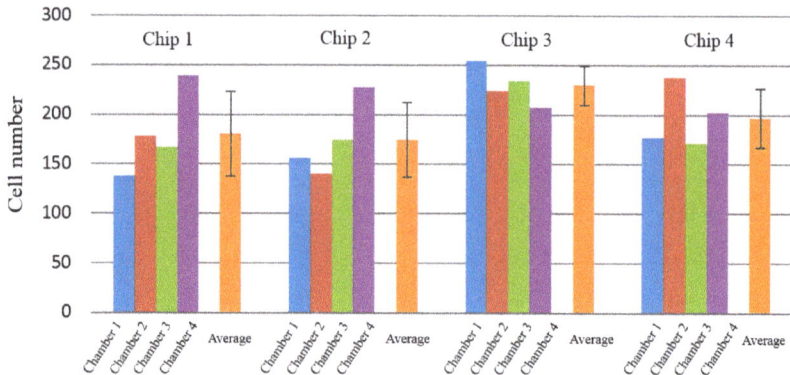

Figure 5. rBMSCs number uniformity in four microfluidic chamber.

In the microfluidic system, the shear stress is generated by the fluid flow injection and stimulates rBMCs. This concept is based on the laminar flow Navier–Stokes theory, since the flow in the microfluidic chamber is laminar, Newtonian flow, and incompressible. The shear stress in the microfluidic chamber surface can be estimated as $\tau_{wall} = \dfrac{6\mu Q}{wh^2}$. Where μ is the fluid dynamic viscosity, Q is the fluid injection flow rate, h is the chamber height, and w is chamber width. Thus, the mechanical shear stress generated on the cell chamber is proportionally relative to the flow rate. For a Y-shaped splitter with equal microchannel width delivering an equal flow rate into each cell chamber, the shear stress generating in each microfluidic chamber is the same. With a different Y-shaped microchannel width, the shear stress generated in each cell chamber will be different. This provides benefits in studying different shear stress effects on a single chip, and its cell culture stability within different microchannels will require further characterization. Figure 6 shows the microfluidic device with 300, 400, 500, and 600 µm microchannel width connecting to cell Chambers 1, 2, 3, and 4, respectively. rBMSCs cell culture results show that the cell number inside Chambers 1, 2, 3, and 4 were measured as 276, 337, 428, and 438, respectively. Since the cell number inside the chamber is related to the flow rate injected into the microchamber and the fluid flow split into each microchannel is based on the hydraulic resistance of each microchannel network. Based on the hydraulic resistance, smaller microchannel width chambers exhibit higher hydraulic resistance than the larger microchannel width chambers. Therefore, an increased cell number tendency shows with the microchannel width. The cell number in Chamber 4 is 1.6 times higher than Chamber 1. However, small cell number difference such as cell number in Chambers 3 and 4 may still be observed due to the chamber-to-chamber variation. In the previous rBMSCs number uniformity test with the same microchannel width (Figure 5), the average cell number variation between each chamber-to-chamber event is 17.2%.

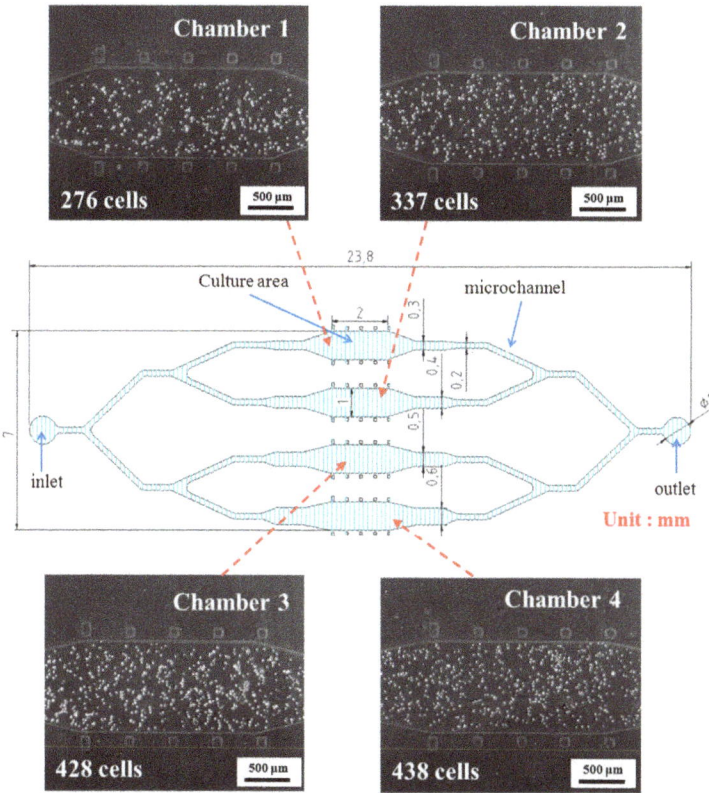

Figure 6. rBMSCs fluid flow shear stress microfluidic chip with different microchannel widths.

3.2. Induction of rBMSCs Differentiation into Neuronal Cells by IBMX Stimulation

To ensure good cell culture and differential stability, we used a microfluidic device with identical microchannel width for the IBMX chemical and physical shear stress stimulation experiments rBMSCs. The rBMSCs were cultured until the cell number approached 400 cells inside each microfluidic chamber after seeding. In a previous investigation [30], we showed that chemical IBMX can efficiently stimulate human placenta-derived stem cells to differentiate into neuronal cells. Neuronal cells can be clearly identified by their cell morphology under optical microscopy. Thus, in this research, we applied IBMX to differentiate rBMSCs into neuronal cells on the microfluidic bioreactor to study the chemical as well as mechanical shear stress effects on rBMSCs. Figure 7 shows the rBMSCs stimulation results under IBMX concentrations from 0.2–0.6 mM and neuronal cell dendrites, as well as condensed and round cell bodies could be observed under optical microscopy, which indicated successful neuronal cell differentiation under IBMX influence. Because the IBMX is toxic to rBMSCs at high concentration. The rBMSCs differentiation ratio is tested under 0.2–0.6 mM gradient concentration, and the rBMSCs differentiation ratio by IBMX is summarized in Figure 7f. From this chart, it can be shown that the neuronal cell differentiation ratio increased with increasing IBMX concentrations. The differentiation ratio reached a plateau at around 0.5–0.6 mM IBMX.

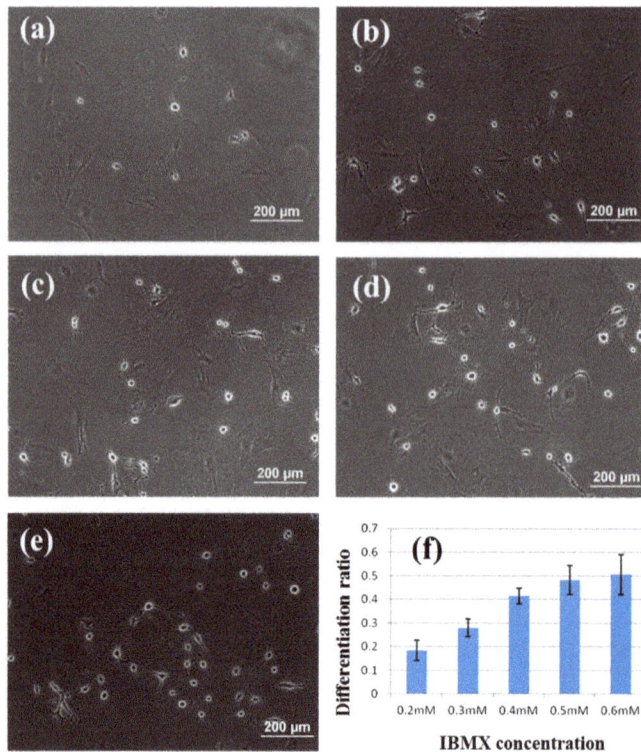

Figure 7. rBMSCs under (**a**) 0.2 mM, (**b**) 0.3 mM, (**c**) 0.4 mM, (**d**) 0.5 mM, and (**e**) 0.6 mM IBMX stimulation. (**f**) Neuronal cell differentiation ratio under various IBMX concentration. Y-axis indicates neuronal cell differentiation ratio and X-axis indicates the IBMX concentration (mM).

3.3. Effects of Fluid Flow Shear Stress on rBMSCs

In addition to the chemical signals, mechanical force is also an important factor in regulating rBMSCs differentiation. Therefore, in addition to IBMX chemical stimulation, we also studied the mechanical shear stress stimulation on the rBMSCs. In the fluid flow shear stress experiments, we introduced 10 min of fluid flow shear stress stimulation on the rBMSCs with addition of 0.2 mM IBMX reagent. Three different fluid flow shear stress conditions were selected: 0.042 µL/min, 1 µL/min, and 15 µL/min. We find rBMSCs will be washed away when the fluid flow injection flow rate exceeds 60 µL/min. Thus, the highest flow rate of 15 µL/min was selected to ensure cells are well adhered on the microfluidic substrate after fluid-flow shear stress stimulation in our experiment. Using a rectangular microfluidic chamber of dimensions 100 µm height and 1000 µm width, fluid flow shear stress was calculated as 0.0009, 0.022, and 0.33 dynes/cm^2 on the rBMSCs. After 10 min of fluid flow shear stress stimulation, the flow rate changed to 0.042 µL/min to culture rBMSCs and we observed neural cell differentiation at 1, 2, 6, and 10 h.

Figure 8a shows the immunocytochemistry images of rBMSCs stained with anti-neuron-specific enolase antibody. The rBMSCs were subject to 0.042, 1, and 15 µL/min fluid flow shear stress stimulations for 10 min and cultured for 1, 2, 6, and 10 h respectively before immunocytochemistry. From the figure, it can be observed that relatively more neuronal cells were found in the microfluidic chamber at higher fluid flow shear stress conditions. Figure 8b summarizes the neuronal cell differentiation ratio. The neuronal cell differentiation ratio is defined by the number of immunocytochemistry-positive cells divided by the total rBMSCs number in the microfluidic chamber at each measurement time. At 0.042 μL/min, the neuronal cell differentiation ratio increased from 0.31 ± 0.02 at 1 h and 0.35 ± 0.04 at 2 h to 0.37 ± 0.06 at 6 h and then decreased to 0.22 ± 0.03 at 10 h. This indicates that rBMSCs start to differentiate into neuronal cells in the first hour and continue differentiating into neuronal cells over the next few hours. Similar differentiation tendencies were also observed at higher fluid flow shear stress conditions of 1 and 15 μL/min. By increasing fluid flow shear stress to 1 μL/min, the neuronal cell differentiation ratio also increased from 0.33 ± 0.03 (1 h) to 0.60 ± 0.05 (2 h) and then decreased to 0.38 ± 0.08 (6 h) and 0.30 ± 0.06 (10 h). At the highest fluid flow condition of 15 μL/min, neuronal cell differentiation ratio increased from 0.57 ± 0.04 to 0.75 ± 0.09 and decreased to 0.53 ± 0.02 for 1, 2, and 6 h, respectively.

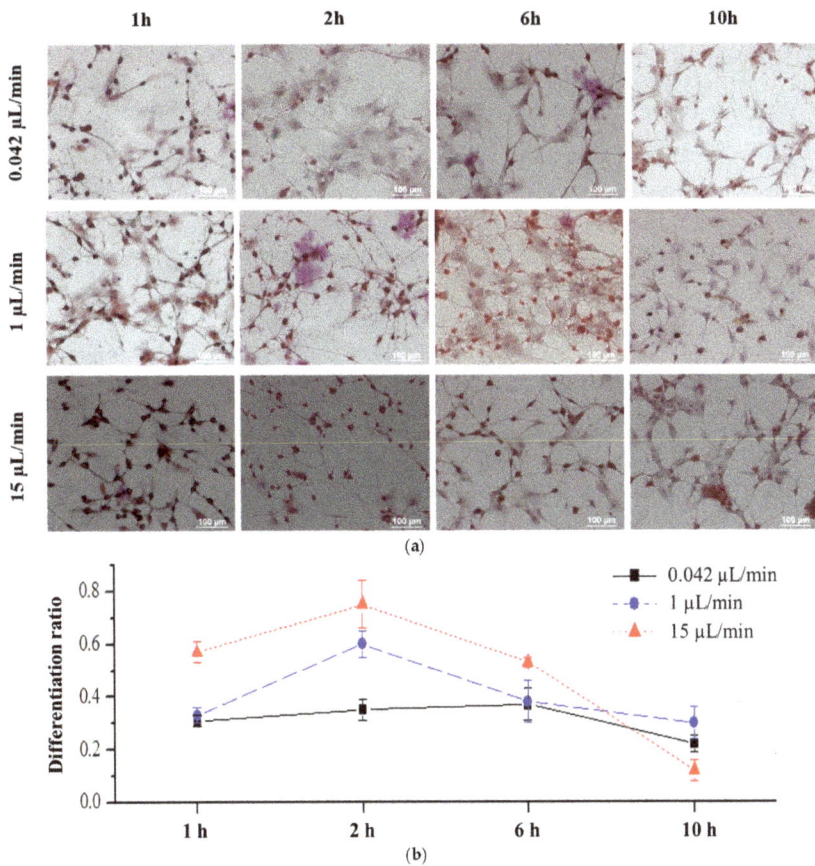

Figure 8. (a) Immunocytochemistry staining images captured from 1, 2, 6, and 10 h after 10 min of 0.042, 1, and 15 μL/min fluid flow shear stress simulation; (b) Neuronal cell differentiation ratio under 10 min of 0.042, 1, and 15 μL/min fluid flow shear stress simulation

rBMSCs neural cell differentiation ratio enhancement were found at increased fluid flow shear stress condition. Highest differentiation ratio of 0.75 were observed at 15 μL/min, two hours condition which is over 2–3 times the neural cell differentiation ratio enhancement than the lowest flow rate condition (neural cell differentiation ratio: 0.22~0.37). The fluid-flow shear stress also accelerate neural cell differentiation earlier. Highest differentiation ratio condition accelerated from six hours (0.042

μL/min) to two hours (1 and 15 μL/min). Additionally, it is found that the neural differentiation ratio decrease at six and 10 h for the high fluid flow shear stress condition of 1 and 15 μL/min. This neural cell number decline was presumably the result from the cell death. The rBMSCs' differentiated neural cells leads the differentiated neuron death after hours, which results in a neural cell differentiation ratio decrease at six and 10 h. In a previous research demonstration, the IBMX-induced neural cells also showed decreased cell numbers after hours [31]. Rishmanchi [32] also found higher differentiation rate results in faster cell death. This also explains the 15 μL/min fluid flow shear stress condition showing a faster differentiation ratio decline in six and 10 h. Regarding the reason of acceleration of neuronal differentiation, there are several factors, such as micro-environment, small molecule, and genetic manipulation were mentioned. As to the micro-environment cues, Akhavan *et al.* reported accelerated differentiation of neural stem cells cultivated on ginseng-reduced graphene oxide sheets [33]. It is addressed that a three-dimensional collagen-hyaluronan matrix could enhance neuronal differentiation [34]. Francis *et al.* demonstrated that preconditioning the human embryonic stem cell with hypoxic environment and its derived cytoprotective phenotype may promote the neural differentiation and enhanced cell survival rates [35]. Chao *et al.* found that using 2D thin film scaffolds composed of biocompatible polymer grafted carbon nanotubes could promote neuron differentiation [36]. These results suggest that environmental cues participate in enhanced neuronal differentiation probably through the cell matrix proteins adsorption and cell attachment and result in the cell physiological responses. Regarding to the small molecules, it is found that higher iron concentration can drastically accelerate the motor neuron differentiation from human embryonic stem cells (hESCs) [37]. A small chemical, called KHS 101, was also found have the ability to accelerate the neuronal differentiation by interaction with TACC3 protein and force the neural progenitor cells to exit the cell cycle [38]. Many reports mentioned neuronal differentiation acceleration are focused on genetic manipulation, such as expression of an extra copy of IκB kinase α blocks self-renewal and accelerates the differentiation of NPCs [39]. The suppression of p53 either by the addition of anti-sense oligonucleotides to culture medium or by the culture of neurons from tumor suppressor protein p53(−/−) mice accelerated their differentiation [40].

In terms of physical stimulation, there is growing evidence suggesting that physical and mechanical stresses, in addition to soluble molecules, may direct cell fate. Stress generated by physical stimulation is known to affect the structure, composition, and function of living tissues [41–43], probably through alterations in the extracellular matrix [44–46]. Chung *et al.* demonstrated a gradient-generating microfluidic device that could be a platform for minimizing autocrine and paracrine signaling and optimizing proliferation and differentiation of neural stem cells in culture [47]. Chowdhury *et al.* showed that a local cyclic stress through focal adhesions induced spreading in mouse embryonic stem cells but not in mouse embryonic stem-cell-differentiated cells. The applied stress also led to gene expression of oct3/4 down-regulation in mES cells [48]. A series researches done by Chien demonstrate the shear stress are able to convert mechanical stimuli into intracellular signals that affect cellular functions of endothelial cells such as proliferation, apoptosis, migration, as well as gene expression. The cytoskeleton provides a structural framework for the endothelial cells to transmit mechanical forces between cell surfaces and cytosol [49–51]. We previously demonstrate the shear stress could accelerate the neuronal differentiation in human placenta-derived multipotent cells [52]. Here we also obtain similar results by utilizing rat bone marrow stromal cells, indicating the shear stress-stimulated neuronal differentiation is a universal phenomenon and may be utilized to improve the effectiveness of human neural precursor transplantation therapies in the future.

4. Conclusions

Microfluidic devices provide advantages in low sample consumption, multiplex, high integration, and good environmental regulation. As an alternative to conventional bioreactors, microfluidic devices are a simple and effective platform for stem cell investigations. In this study, we used a microfluidic device to deliver chemical IBMX stimuli and physical fluid flow shear stress to stimulate rBMSCs in four parallel microfluidic chamber formats. The microfluidic bioreactor with the same microchannel width showed good stability with an average RSD% of 17.3% between microfluidic chambers and an average RSD% of 18.9% between microfluidic device events. We showed that rBMSCs can be differentiated into neuronal cells by IBMX. Enhanced neuronal cell differentiation behavior was observed when rBMSCs were subjected to mechanical shear stress created by microfluidic flow injection. With increased fluid flow shear stress, more rBMSCs differentiated into neuronal cells. The highest neuronal cell differentiation ratio of 0.75 was achieved at the highest fluid flow rate condition of 15 µL/min. Our findings identify a significant role afforded by shear stress in neuronal differentiation. The impact of the mechanical forces on neuronal differentiation allows us to better understand stem cell responses to chemical and mechanical manipulations and may contribute to the area of tissue engineering in the future.

Acknowledgments: Acknowledgments: The authors would like to thank National Science Council, Taiwan (Grant # MOST 104-2221-E-008 -080 and NCU-CGH joint research program 103CGH-NCU-A2) financially supporting this project.

Author Contributions: Author Contributions: Chia-Wen Tsao and Yu-Che Cheng conceived and designed the experiments, contributed reagents/materials/analysis tools, analyzed the data and wrote the paper; Jhih-Hao Cheng performed the experiments.

Conflicts of Interest: Conflicts of Interest: The authors declare no conflict of interest.

References

1. Mehling, M.; Tay, S. Microfluidic cell culture. *Curr. Opin. Biotech.* **2014**, *25*, 95–102. [CrossRef] [PubMed]
2. Davidsson, R.; Boketoft, A.; Bristulf, J.; Kotarsky, K.; Olde, B.; Owman, C.; Bengtsson, M.; Laurell, T.; Emneus, J. Developments toward a microfluidic system for long-term monitoring of dynamic cellular events in immobilized human cells. *Anal. Chem.* **2004**, *76*, 4715–4720. [CrossRef] [PubMed]
3. Mellors, J.S.; Jorabchi, K.; Smith, L.M.; Ramsey, J.M. Integrated microfluidic device for automated single cell analysis using electrophoretic separation and electrospray ionization mass spectrometry. *Anal. Chem.* **2010**, *82*, 967–973. [CrossRef] [PubMed]
4. Mazutis, L.; Gilbert, J.; Ung, W.L.; Weitz, D.A.; Griffiths, A.D.; Heyman, J.A. Single-cell analysis and sorting using droplet-based microfluidics. *Nat. Protoc.* **2013**, *8*, 870–891. [CrossRef] [PubMed]
5. Park, T.H.; Shuler, M.L. Integration of cell culture and microfabrication technology. *Biotechnol. Progress* **2003**, *19*, 243–253. [CrossRef] [PubMed]
6. Breslauer, D.N.; Lee, P.J.; Lee, L.P. Microfluidics-based systems biology. *Mol. Biosyst.* **2006**, *2*, 97–112. [CrossRef] [PubMed]
7. McDonald, J.C.; Duffy, D.C.; Anderson, J.R.; Chiu, D.T.; Wu, H.K.; Schueller, O.J.A.; Whitesides, G.M. Fabrication of microfluidic systems in poly(dimethylsiloxane). *Electrophoresis* **2000**, *21*, 27–40. [CrossRef]
8. Melin, J.; Quake, S.R. Microfluidic large-scale integration: The evolution of design rules for biological automation. *Annu. Rev. Biophys. Biomol.* **2007**, *36*, 213–231. [CrossRef] [PubMed]
9. Toh, Y.C.; Zhang, C.; Zhang, J.; Khong, Y.M.; Chang, S.; Samper, V.D.; van Noort, D.; Hutmacher, D.W.; Yu, H. A novel 3D mammalian cell perfusion-culture system in microfluidic channels. *Lab Chip* **2007**, *7*, 302–309. [CrossRef] [PubMed]
10. Reardon, S. "Organs-on-chips" go mainstream. *Nature* **2015**, *523*, 266. [CrossRef] [PubMed]
11. Bhatia, S.N.; Ingber, D.E. Microfluidic organs-on-chips. *Nat. Biotechnol.* **2014**, *32*, 760–772. [CrossRef] [PubMed]
12. Ashton, R.S.; Keung, A.J.; Peltier, J.; Schaffer, D.V. Progress and prospects for stem cell engineering. *Annu. Rev. Chem. Biomol.* **2011**, *2*, 479–502. [CrossRef] [PubMed]
13. Steward, A.J.; Thorpe, S.D.; Vinardell, T.; Buckley, C.T.; Wagner, D.R.; Kelly, D.J. Cell-matrix interactions regulate mesenchymal stem cell response to hydrostatic pressure. *Acta Biomater.* **2012**, *8*, 2153–2159. [CrossRef] [PubMed]
14. Miyanishi, K.; Trindade, M.C.D.; Lindsey, D.P.; Beaupre, G.S.; Carter, D.R.; Goodman, S.B.; Schurman, D.J.; Smith, R.L. Effects of hydrostatic pressure and transforming growth factor-β3 on adult human mesenchymal stem cell chondrogenesis *in vitro*. *Tissue Eng.* **2006**, *12*, 1419–1428. [CrossRef] [PubMed]

15. Liu, Y.R.; Buckley, C.T.; Mulhall, K.J.; Kelly, D.J. Combining BMP-6, TGF-β3 and hydrostatic pressure stimulation enhances the functional development of cartilage tissues engineered using human infrapatellar fat pad derived stem cells. *Biomater. Sci.* **2013**, *1*, 745–752. [CrossRef]

16. Kim, D.; Heo, S.J.; Kim, S.H.; Shin, J.; Park, S.; Shin, J.W. Shear stress magnitude is critical in regulating the differentiation of mesenchymal stem cells even with endothelial growth medium. *Biotechnol. Lett.* **2011**, *33*, 2351–2359. [CrossRef] [PubMed]

17. Nikmanesh, M.; Shi, Z.D.; Tarbell, J.M. Heparan sulfate proteoglycan mediates shear stress-induced endothelial gene expression in mouse embryonic stem cell-derived endothelial cells. *Biotechnol. Bioeng.* **2012**, *109*, 583–594. [CrossRef] [PubMed]

18. Kearney, E.M.; Farrell, E.; Prendergast, P.J.; Campbell, V.A. Tensile strain as a regulator of mesenchymal stem cell osteogenesis. *Ann. Biomed. Eng.* **2010**, *38*, 1767–1779. [CrossRef] [PubMed]

19. Sumanasinghe, R.D.; Osborne, J.A.; Loboa, E.G. Mesenchymal stem cell-seeded collagen matrices for bone repair: Effects of cyclic tensile strain, cell density, and media conditions on matrix contraction *in vitro*. *J. Biomed. Mater. Res. A* **2009**, *88A*, 778–786. [CrossRef] [PubMed]

20. Wu, H.W.; Lin, C.C.; Lee, G.B. Stem cells in microfluidics. *Biomicrofluidics* **2011**, *5*, 013401. [CrossRef] [PubMed]

21. Van Noort, D.; Ong, S.M.; Zhang, C.; Zhang, S.F.; Arooz, T.; Yu, H. Stem cells in microfluidics. *Biotechnol. Prog.* **2009**, *25*, 52–60. [CrossRef] [PubMed]

22. Jeon, N.L.; Dertinger, S.K.W.; Chiu, D.T.; Choi, I.S.; Stroock, A.D.; Whitesides, G.M. Generation of solution and surface gradients using microfluidic systems. *Langmuir* **2000**, *16*, 8311–8316. [CrossRef]

23. Thorpe, S.D.; Buckley, C.T.; Vinardell, T.; O'Brien, F.J.; Campbell, V.A.; Kelly, D.J. The response of bone marrow-derived mesenchymal stem cells to dynamic compression following TGF-β3 induced chondrogenic differentiation. *Ann. Biomed. Eng.* **2010**, *38*, 2896–2909. [CrossRef] [PubMed]

24. Breen, L.T.; McHugh, P.E.; McCormack, B.A.; Muir, G.; Quinlan, N.J.; Heraty, K.B.; Murphy, B.P. Development of a novel bioreactor to apply shear stress and tensile strain simultaneously to cell monolayers. *Rev. Sci. Instrum.* **2006**, *77*, 104301. [CrossRef]

25. Blackman, B.R.; Garcia-Cardena, G.; Gimbrone, M.A. A new *in vitro* model to evaluate differential responses of endothelial cells to simulated arterial shear stress waveforms. *J. Biomech. Eng.* **2002**, *124*, 397–407. [CrossRef] [PubMed]

26. Park, S.W.; Byun, D.; Bae, Y.M.; Choi, B.H.; Park, S.H.; Kim, B.; Cho, S.I. Effects of fluid flow on voltage-dependent calcium channels in rat vascular myocytes: Fluid flow as a shear stress and a source of artifacts during patch-clamp studies. *Biochem. Biophys. Res. Commun* **2007**, *358*, 1021–1027. [CrossRef] [PubMed]

27. Korin, N.; Bransky, A.; Dinnar, U.; Levenberg, S. Periodic "flow-stop" perfusion microchannel bioreactors for mammalian and human embryonic stem cell long-term culture. *Biomed. Microdevices* **2009**, *11*, 87–94. [CrossRef] [PubMed]

28. Sim, W.Y.; Park, S.W.; Park, S.H.; Min, B.H.; Park, S.R.; Yang, S.S. A pneumatic micro cell chip for the differentiation of human mesenchymal stem cells under mechanical stimulation. *Lab Chip* **2007**, *7*, 1775–1782. [CrossRef] [PubMed]

29. Tsao, C.W.; Liu, J.; Devoe, D.L. Droplet formation from hydrodynamically coupled capillaries for parallel microfluidic contact spotting. *J. Micromech. Microeng.* **2008**, *18*, 025013. [CrossRef]

30. Chien, C.C.; Yen, B.L.; Lee, F.K.; Lai, T.H.; Chen, Y.C.; Chan, S.H.; Huang, H.I. *In vitro* differentiation of human placenta-derived multipotent cells into hepatocyte-like cells. *Stem Cells* **2006**, *24*, 1759–1768. [CrossRef] [PubMed]

31. Yen, B.L.; Chien, C.C.; Chen, Y.C.; Chen, J.T.; Huang, J.S.; Lee, K.; Huang, H.I. Placenta-derived multipotent cells differentiate into neuronal and glial cells *in vitro*. *Tissue Eng. A* **2008**, *14*, 9–17. [CrossRef] [PubMed]

32. Rismanchi, N.; Floyd, C.L.; Berman, R.F.; Lyeth, B.G. Cell death and long-term maintenance of neuron-like state after differentiation of rat bone marrow stromal cells: A comparison of protocols. *Brain Res.* **2003**, *991*, 46–55. [CrossRef] [PubMed]

33. Akhavan, O.; Ghaderi, E.; Abouei, E.; Hatamie, S.; Ghasemi, E. Accelerated differentiation of neural stem cells into neurons on ginseng-reduced graphene oxide sheets. *Carbon* **2014**, *66*, 395–406. [CrossRef]

34. Brannvall, K.; Bergman, K.; Wallenquist, U.; Svahn, S.; Bowden, T.; Hilborn, J.; Forsberg-Nilsson, K. Enhanced neuronal differentiation in a three-dimensional collagen-hyaluronan matrix. *J. Neurosci. Res.* **2007**, *85*, 2138–2146. [CrossRef] [PubMed]

35. Francis, K.R.; Wei, L. Human embryonic stem cell neural differentiation and enhanced cell survival promoted by hypoxic preconditioning. *Cell Death Dis.* **2010**, *1*, e22. [CrossRef] [PubMed]

36. Chao, T.I.; Xiang, S.; Chen, C.S.; Chin, W.C.; Nelson, A.J.; Wang, C.; Lu, J. Carbon nanotubes promote neuron differentiation from human embryonic stem cells. *Biochem. Biophys. Res. Commun.* **2009**, *384*, 426–430. [CrossRef] [PubMed]

37. Lu, D.; Chen, E.Y.; Lee, P.; Wang, Y.C.; Ching, W.; Markey, C.; Gulstrom, C.; Chen, L.C.; Nguyen, T.; Chin, W.C. Accelerated neuronal differentiation toward motor neuron lineage from human embryonic stem cell line (H9). *Tissue Eng. C Methods* **2015**, *21*, 242–252. [CrossRef] [PubMed]

38. Wurdak, H.; Zhu, S.; Min, K.H.; Aimone, L.; Lairson, L.L.; Watson, J.; Chopiuk, G.; Demas, J.; Charette, B.; Halder, R.; *et al.* A small molecule accelerates neuronal differentiation in the adult rat. *Proc. Natl. Acad. Sci. USA* **2010**, *107*, 16542–16547. [CrossRef] [PubMed]

39. Khoshnan, A.; Patterson, P.H. Elevated ikkalpha accelerates the differentiation of human neuronal progenitor cells and induces MeCP2-dependent BDNF expression. *PLoS ONE* **2012**, *7*, e41794. [CrossRef] [PubMed]

40. Ferreira, A.; Kosik, K.S. Accelerated neuronal differentiation induced by p53 suppression. *J. Cell Sci.* **1996**, *109*, 1509–1516. [PubMed]

41. Masuda, T.; Takahashi, I.; Anada, T.; Arai, F.; Fukuda, T.; Takano-Yamamoto, T.; Suzuki, O. Development of a cell culture system loading cyclic mechanical strain to chondrogenic cells. *J. Biotechnol.* **2008**, *133*, 231–238. [CrossRef] [PubMed]

42. Knothe, U.R.; Dolejs, S.; Matthew Miller, R.; Knothe Tate, M.L. Effects of mechanical loading patterns, bone graft, and proximity to periosteum on bone defect healing. *J. Biomech.* **2010**, *43*, 2728–2737. [CrossRef] [PubMed]

43. Nagatomi, J.; Arulanandam, B.P.; Metzger, D.W.; Meunier, A.; Bizios, R. Effects of cyclic pressure on bone marrow cell cultures. *J. Biomech. Eng.* **2002**, *124*, 308–314. [PubMed]

44. Schumann, D.; Kujat, R.; Nerlich, M.; Angele, P. Mechanobiological conditioning of stem cells for cartilage tissue engineering. *Biomed. Mater. Eng.* **2006**, *16*, S37–S52. [PubMed]

45. Zioupos, P.; Cook, R.B.; Hutchinson, J.R. Some basic relationships between density values in cancellous and cortical bone. *J. Biomech.* **2008**, *41*, 1961–1968. [CrossRef] [PubMed]

46. Lee, T.C.; O'Brien, F.J.; Gunnlaugsson, T.; Parkesh, R.; Taylor, D. Microdamage and bone mechanobiology. *Technol. Health Care* **2006**, *14*, 359–365. [PubMed]

47. Chung, B.G.; Flanagan, L.A.; Rhee, S.W.; Schwartz, P.H.; Lee, A.P.; Monuki, E.S.; Jeon, N.L. Human neural stem cell growth and differentiation in a gradient-generating microfluidic device. *Lab Chip* **2005**, *5*, 401–406. [CrossRef] [PubMed]

48. Chowdhury, F.; Na, S.; Li, D.; Poh, Y.-C.; Tanaka, T.S.; Wang, F.; Wang, N. Material properties of the cell dictate stress-induced spreading and differentiation in embryonic stem cells. *Nat. Mater.* **2010**, *9*, 82–88. [CrossRef] [PubMed]

49. Lin, K.; Hsu, P.P.; Chen, B.P.; Yuan, S.; Usami, S.; Shyy, J.Y.; Li, Y.S.; Chien, S. Molecular mechanism of endothelial growth arrest by laminar shear stress. *Proc. Natl. Acad. Sci. USA* **2000**, *97*, 9385–9389. [CrossRef] [PubMed]

50. Jalali, S.; del Pozo, M.A.; Chen, K.; Miao, H.; Li, Y.; Schwartz, M.A.; Shyy, J.Y.; Chien, S. Integrin-mediated mechanotransduction requires its dynamic interaction with specific extracellular matrix (ECM) ligands. *Proc. Natl. Acad. Sci. USA* **2001**, *98*, 1042–1046. [CrossRef] [PubMed]

51. Wang, K.C.; Garmire, L.X.; Young, A.; Nguyen, P.; Trinh, A.; Subramaniam, S.; Wang, N.; Shyy, J.Y.; Li, Y.S.; Chien, S. Role of microRNA-23b in flow-regulation of Rb phosphorylation and endothelial cell growth. *Proc. Natl. Acad. Sci. USA* **2010**, *107*, 3234–3239. [CrossRef] [PubMed]

52. Cheng, Y.-C.; Tsao, C.-W.; Chiang, M.-Z.; Chung, C.-A.; Chien, C.-C.; Hu, W.-H.; Ruaan, R.-C.; Li, C. Microfluidic platform for human placenta-derived multipotent stem cells culture and applied for enhanced neuronal differentiation. *Microfluid. Nanofluid.* **2015**, *18*, 587–598. [CrossRef]

micromachines

MDPI

Article

Liquid Gradient Refractive Index Microlens for Dynamically Adjusting the Beam Focusing

Zichun Le *, Yunli Sun and Ying Du

College of Sciences, Zhejiang University of Technology, Hangzhou 310023, China;
sy1785809@sina.com (Y.S.); duying@zjut.edu.cn (Y.D.)
* Correspondance: lzc@zjut.edu.cn; Tel.: +086-571-85290552

Academic Editors: Shih-Kang Fan and Nam-Trung Nguyen
Received: 23 September 2015; Accepted: 7 December 2015; Published: 10 December 2015

Abstract: An in-plane liquid gradient index (L-GRIN) microlens is designed for dynamically adjusting the beam focusing. The ethylene glycol solution (core liquid) with de-ionized (DI) water (cladding liquid) is co-injected into the lens chamber to form a gradient refractive index profile. The influences of the diffusion coefficient, mass fraction of ethylene glycol and flow rate of liquids on the refractive index profile of L-GRIN microlens are analyzed, and the finite element method and ray tracing method are used to simulate the convection-diffusion process and beam focusing process, which is helpful for the prediction of focusing effects and manipulation of the device. It is found that not only the focal length but the focal spot of the output beam can be adjusted by the diffusion coefficient, mass fraction and flow rate of liquids. The focal length of the microlens varies from 942 to 11 µm when the mass fraction of the ethylene glycol solution varies from 0.05 to 0.4, and the focal length changes from 127.1 to 8 µm by varying the flow rate of the core liquid from 0.5×10^3 to 5×10^3 pL/s when there is no slip between the core and cladding inlet. The multiple adjustable microlens with a simple planar microfluidic structure can be used in integrated optics and lab-on-chip systems.

Keywords: in-plane liquid gradient index (L-GRIN) microlens; optofluidics waveguide; adjustable focal length; convection-diffusion process; beam focusing; finite element method

1. Introduction

Tunable microlenses are widely used in microfluidic or lab-on-chip systems [1] due to their fine tuning of the light. A tunable microlens adaptively reshapes the input light including adjusting its focal position, intensity, beam profile, and even propagation direction [2]. They can also be used for ordinary coupling, collimating and focusing, and in wide range of lab-on-a-chip applications such as flow cytometry [3–5], on-chip optical tweezers and single molecule detection [6,7].

Several kinds of tunable microlenses are developed for providing adaptive focusing with adjustable curved refractive surfaces [2], such as a micromeniscus surface actuated by electro-wetting [8], liquid microlenses based on the dielectrophoretic effect [9–11], microlenses based on the thermal effect [12], or hydraulically actuated polydimethylsiloxane (PDMS) membranes [13,14]. Among them, electro-wetting and dielectrophoresis lenses are electrically driven, while other microlenses whose focal lengths are variable are either based on pressure-induced systems using an elastic membrane, or on thermal effect. A mechanical lens whose focal length is controlled by pumping liquid in and out of the lens chamber caused great interest. The operation mechanism of this lens is simple because it only requires a fluid pumping system. There are two working mechanisms which are often used to implement these tunable liquid microlenses. One is to achieve the refraction of light at curved, optically smooth, liquid-liquid interfaces between two flowing fluids by controlling laminar flows. Recently, the Whitesides group has developed a dynamically reconfigurable liquid-liquid lens [15] with the convex shape of liquid-liquid interfaces in a microfluidic expansion chamber. The curvatures in the liquid-liquid interfaces were controlled by adjusting hydrodynamic flow conditions, and then adjustable focal length was produced. However, in the liquid-liquid lens, a higher speed of the laminar flows is necessary, which means large amounts of fluid have to be supplied discontinuously for keeping the microlens working. Utilization of so-called liquid gradient refractive index (L-GRIN) microlenses [16] is the other approach and also the preferable way to achieve a tunable microlens. Similarly, with the traditional solid self-focusing lens, the refractive index of the L-GRIN microlens increases continuously from the cladding region to the core optical axis. However, a gradient refractive index of the L-GRIN microlens is achieved by dynamically adjusting hydrodynamic flow conditions rather than by changing the geometrical structure of the lens for the traditional one. In contrast to the liquid-liquid interface-based microlens, the L-GRIN lens operates through the diffusion in multiple flows rather than by relying on a clearly defined, curved liquid-liquid interface. Therefore, the flow consumption rate is much lower in the L-GRIN lens [17]. These lenses focus a light beam through a liquid medium with a two-dimensional (2D) refractive index gradient. The two-dimensional (2D) refractive index profile and, subsequently, the focal length of the L-GRIN lens can be tuned by changing the ratio of the flow rates of the two inlets [17]. However, most L-GRIN microlenses are only capable of out-of-plane focusing. In-plane-focusing tunable microlens are, however, proven to be preferable in integrative devices, which can be fabricated and seamlessly integrated with other on-chip fluidic and optical components such as a lab-on-a-chip lasers [18] and optical waveguides [19]. In particular, it has inspired the creation of a variety of innovative devices controlled by flow rates and liquid compositions, and variable light focusing was measured and shown, but the quantitative relation between liquid parameters and focal length has not been demonstrated clearly. It will lead to the unpredictability of the focusing effect in the device.

In this paper, an in-plane tunable L-GRIN microlens was designed for dynamically adjusting the beam focusing, which can be more readily integrated for lab-on-a-chip applications. The convection-diffusion process of liquids in the microfluidic chamber is firstly simulated with the finite element method (FEM) when a high-refractive-index solution is injected side-by-side into a low-refractive-index solution. During the convection and diffusing process, the diffusion profile of the liquids, and hence the refractive index profile within the L-GRIN microlens, varies with the hydrodynamic flow conditions. Thus, the refractive index profile of the L-GRIN microlens is calculated and discussed numerically under different flow conditions, and the beam transmission and focusing process in the L-GRIN microlens is simulated using the ray tracing method. The effects of the diffusion coefficient, mass fraction and flow rate of liquids on the input beam focusing effect, including the focal length and the size of the focal spot, were demonstrated. Therefore, an in-plane tunable beam focusing is achieved using a simple planar microfluidic structure; in addition, with the simulation of the convection and diffusing process, the ray tracing method gives us approaches to predict the focusing effect of an L-GRIN microlens.

2. Structure and Principle of L-GRIN Micrlens

The schematic of the L-GRIN microlens designed is shown in Figure 1. The structure of the microlens as shown in Figure 1a consists of the microfluid chamber, core inlet, cladding inlet and outlet. The core liquid and the cladding liquid are injected through the core inlet and cladding inlet, respectively, which flow out through the outlet. H_{in_core}, W_{in_core}, H_{in_clad}, W_{in_clad}, H_{out}, W_{out} represent the height and width of the core inlet, cladding inlet and outlet, respectively. The main part of the L-GRIN microlens is a micro-cylindrical chamber, where the diffusion and convection process of liquids occurs and the gradient refractive index profile appears. The *xoy* cross-sectional view of the L-GRIN microlens together with the substrate is also illustrated in Figure 1a, where d_{core} which is designed to be 50 μm stands for the diameter of the core inlet. Similarly, d_{clad} is the diameter of the cladding inlet and is designed to be 150 μm. In the chamber, ethylene glycolsolution (core liquid) is injected side-by-side into de-ionized (DI) water (cladding liquid) from the same direction as shown in Figure 1b. On account of the well-known manner in which the refractive index of liquids varies with the wavelength of incident light, in our simulation, the wavelength of incident light is set to be 500nm without considering the dispersion of light. The diffusion and convection process between ethylene glycol solution and DI water results in a 2D refractive index profile in the lens chamber, as shown in the left graph of Figure 1c.An axis-symmetric refractive index gradient is accordingly achieved in the *xoy* plane and the simulation result is given in the right graph of Figure 1c, which shows that the refractive index of the L-GRIN microlens increases continuously from the cladding region to the core optical axis. The maximum refractive index comes near the inlet, and under the influence of diffusion, the refractive index decreases along the flowing direction of the liquids.

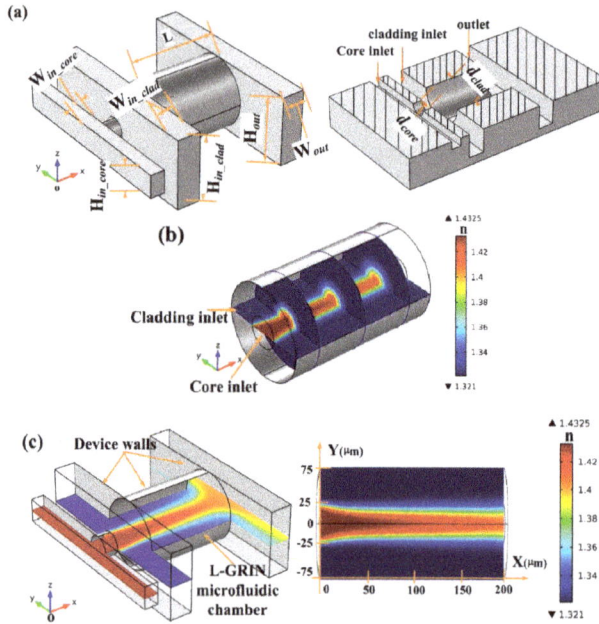

Figure 1. (a) Structure diagram of the tunable L-GRIN microlens (left graph), in which H_{in_core} = 50 μm, W_{in_core} = 30 μm, H_{in_clad} = 150 μm, W_{in_clad} = 50 μm, H_{out} = 150 μm, W_{out} = 50 μm. The *xoy* cross-sectional view of the L-GRIN microlens together with the substrate (right graph), in which d_{core} = 50 μm, d_{clad} = 150 μm; (b)The simplified model of the microfluidic chamber; (c) Simulated refractive index profile in the microfluidic chamber (left graph) and the refractive index profile in the *xoy* plane (right graph).

To simulate and optimize the refractive index profile and light propagation in the L-GRIN microlens under different conditions, the FEM and optical ray tracing method are adopted. The refractive index profile can be calculated by simulating the diffusion and convection process of liquids in the microfluidic chamber. When the refractive index of liquids is mainly affected by diffusion, the process can be described by Fick's second law:

$$\frac{\partial C}{\partial t} = D\nabla^2 C \qquad (1)$$

In which C is the solution concentration and t is the diffusion time. D represents the diffusion coefficient that may determine the diffusion speed between the core liquid and the cladding liquid. As shown in Equation (1), the solution concentration changes with time and space. However, in the microfluidic chamber, the diffusion and the convection have a combined effect on the refractive index profile of the L-GRIN microlens. The convective-diffusive process in the chamber is therefore rewritten as [20]:

$$\frac{\partial C}{\partial t} = D\nabla^2 C - U\nabla C \tag{2}$$

where U is defined to be the average flow velocity in the chamber. For a steady state flow, the distribution of concentration in the device does not vary with time, which means that $\partial C/\partial t = 0$. Considering the symmetry of the lens structure, the convective-diffusive process can be simplified as:

$$D\left(\frac{\partial^2 C}{\partial x^2} + \frac{\partial^2 C}{\partial y^2}\right) - U\left(\frac{\partial C}{\partial x} + \frac{\partial C}{\partial y}\right) = 0 \tag{3}$$

The concentration distribution for the full, developed, steady-state flow can be expressed as Equation (4) [21].

$$c'(x',y') = r + \frac{2}{\pi}\sum_{n=1}^{\infty}\frac{\sin(\pi n r)}{n}\cos(n\pi y')\exp[\frac{1}{2}(Pe - \sqrt{Pe^2 + 4n^2\pi^2})x'] \tag{4}$$

In the normalized coordinate system, $x' = x/R$, $y' = y/R$, where R is the radius of the microfluidic chamber. Then $c' = c/C_0$ is the normalized concentration, and C_0 stands for the initial concentration. Pe is called the Peclet number, which is a dimensionless number and represents the relative proportion of convection to diffusion. Pe is proportional to the laminar velocity across the transversal section of the microfluidic chamber and the length of the microfluidic chamber, and Pe is also inversely proportional to the diffusion coefficient. Pe is usually used to describe the coefficient involved in the convection and diffusion process for a common analysis. However, the effect caused by convection and diffusion is analyzed separately in the present paper in order to dynamically reshape the light beam of the L-GRIN microlens.

For the structure we designed, $U = (Q_{core} + Q_{clad})/R^2\pi$ is defined as the average velocity, where Q_{core} is the flow rate of the core liquid and Q_{core} is the flow rate of the cladding liquid; r is the normalized initial boundary ratio between the core and cladding liquids and is expressed as [22]:

$$\alpha\beta^2(r-1)^4 + r(r-1)[r^2 + 3r - \alpha(r-1)(r-4)]\beta - r^4 = 0 \tag{5}$$

where $\alpha = Q_{core}/Q_{clad}$, $\beta = \mu_{core}/\mu_{clad}$, and μ_{core} is the liquid viscosity of the core liquid and μ_{clad} is the liquid viscosity of the cladding liquid. The initial parameters are set as $c'|_{x'=0,0\leqslant y'<r=1} = 1$, $c'|_{x'=0,y'=r} = r$ and $c'|_{x'=0,r\leqslant y'\leqslant 1} = 0$, in which $\frac{\partial c'}{\partial y'}|_{x=\infty,0\leqslant y'\leqslant 1} = 0$ stands for the full mixing state at the outlet and $\frac{\partial c'}{\partial y'}|_{y=1} = 0$ represents the full mixing state at the chamber walls.

Once the liquids for the core and the cladding liquids are determined, the liquid viscosity and diffusion coefficient are considered to be constant. The position of the focal point can be tuned along the x-direction by changing the refractive index profile, which can be achieved by adjusting the flow rate of the core or cladding liquid. According to Equation (2), if the average velocity U is large, the diffusion in the *xoy* plane can be neglected and only the refractive index profile in the *yoz* plane needs to be calculated, where the plane is perpendicular to the liquid flow direction. On the contrary, when the average velocity U is small, the diffusion in both x and y directions has to be discussed. The refractive index profile can be obtained according to the concentration profile [23].

$$n(x',y') = c'_{core}(x',y')n_{core} + c'_{clad}(x',y')n_{clad} \tag{6}$$

where $n(x', y')$ is the refractive index profile of the L-GRIN microlens in the $x'oy'$ plane, and c'_{core} and c'_{clad} are the normalized concentration profile of the core and cladding liquids in the $x'oy'$ plane, respectively. The focusing of the input light beam through the microlens is determined by the refractive index profile.

Because of the decisive effect of the convection-diffusion process on the refractive index profile of the L-GRIN microlens, the average velocity U and the diffusion coefficient D, which are also found to vary with the concentration C and temperature, are important parameters for tunable focusing. For example, the diffusion coefficients between DI water and ethylene glycol are 3.75×10^{-10} and 1.17×10^{-9} m^2/s with different mass fractions of 0.025 and 0.95 for ethylene glycol [24]. In addition, when the temperature varies from 30 to 50 °C, the diffusion coefficient changes from 3.15×10^{-10} to 6.45×10^{-10} m^2/s with a fixed mass fraction of 0.8 for the ethylene glycol [25]. Due to the temperature of liquids usually being stationary during the convection-diffusion process, thediffusion coefficient D, concentration C and average velocity U of liquids are the main influence factors for adjusting the focusing performance of the L-GRIN microlens. It was found by simulations as follows that an adjustable focal length could be achieved under a relatively lower average velocity while the size of the focal spot had to be tuned at a high enough average velocity. Therefore, both the focal length and the focal spot of the output beam can be adjusted by carefully controlling the average velocity of the liquids. Suppose that the DI water and ethylene glycol solution are chosen as the cladding and the core liquids, respectively, and two cases of the average velocity with relative slip and without relative slip between the core and cladding inlet are disscused. The subsequent discussion will be focused on how the adjustment of the output beam will be implemented by adjusting these influence factors.

In order to validate our numerical simulation, the comparison between the observed light propagation in Reference [23] and our simulation results was illustrated in Figure 2. Figure 2a shows light propagation in the optofluidic waveguides under different fluidic conditions, and Figure 2b shows our simulation results at the same fluidic conditions. When driven at a highflow rate of $Q_{clad} = 50$ μL·min^{-1} ($Pe = 0.07$), the light propagates in a straight line and shows no focusing effect. In contrast, at a low flow rate of $Q_{clad} = 0.5$ μL·min^{-1}, the light exhibits curved-ray trajectories and converges repeatedly for different core widths, $r = 1/6, 1/3$ and $1/2$ ($Pe = 0.0005, 0.0007$ and 0.001, respectively) [23].

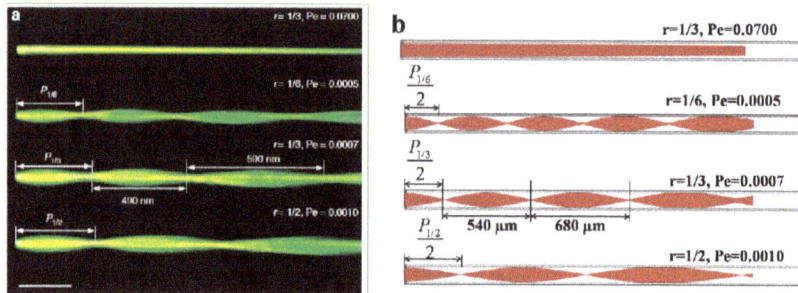

Figure 2. (a) Observed light propagation under different flow conditions [23]. At a high flow rate of $Q_{clad} = 50$ μL·min^{-1}, the light is confined in the core due to the step-index distribution. At $Q_{clad} = 0.5$ μL·min^{-1}, the diffusion-induced gradient of the refractive index causes the light to repeatedly merge. With increasing r, the length of the first section becomes larger, for example $P_{1/6} = 270$ μm, $P_{1/3} = 300$ μm and $P_{1/2} = 340$ μm. The focusing period also increases; for example, the first period for $P_{1/3}$ is 300 μm, the second is extended to 490 μm and the third is 590 μm. (Scale bar equals 300 μm.)(b) Simulated light propagation under the same flow conditions. With increasing r, the length of the first section becomes larger, for example, $P_{1/6} = 300$ μm, $P_{1/3} = 360$ μm and $P_{1/2} = 410$ μm. The focusing period also increases; for example, the first period for $P_{1/3}$ is 360 μm, the second is extended to 540 μm and the third is 680 μm.

Similarly, the length of the first section increases with the core width and decreases almost linearly with the flow rate of the core fluid in both figures. However, when examining the light propagation pattern in detail, it is found that the self-focusing period is chirped such that the focusing period increases. For instance, as shown in Figure 2a, the first period for $P_{1/3}$ is 300 μm, the second is extended to 490 μm, and the third is 590 μm, owing to the diminishing bidirectional gradient contrast downstream as a result of diffusion [23]. Furthermore, the first period for $P_{1/3}$ is 360 μm, the second is extended to 540 μm, and the third is 680 μm in Figure 2b. The small difference in focus length maybe due to the difference between two structures, and the change trends are the same. The match between the experimental results in Reference [23] and our simulation results illustrated that the follow-up simulations could be trusted.

3. Results and Discussion

3.1. TheRefractive Index Profile of the L-GRIN Microlens

In order to form the gradient refractive index profile in the lens chamber, the ethylene glycolsolution (n_{core} = 1.432) and DI water (n_{clad} = 1.332) are co-injected into the lens chamber from the same direction.Once contacting DI water, the ethylene glycol starts to diffuse from the ethylene glycol solution into the DI water. The average velocity U is calculated according to $U = (Q_{core} + Q_{clad})/R^2\pi$. For the purpose of maintaining the same flow rate of the ethylene glycol solution and the DI water (supposingthere is norelative slip between the core and cladding liquids), Q_{clad}/Q_{core} = 8 should be satisfied because the cross-sectional area of the cladding inlet is eight times as large as that of the core inlet. In order to illustrate the influence of the diffusion coefficient D on the refractive index profile of the L-GRIN microlens, the refractive index profiles in the *xoy* plane were simulated at two different diffusion coefficients ($D = 1 \times 10^{-9}$ m^2/s and $D = 4 \times 10^{-10}$ m^2/s) as shown in Figure 3. Figure 3a,b indicate that the variation of the diffusion coefficient resulting from the variation of the environmental conditions such as temperature has a strong influence on the refractive index profile, and the diffusion phenomenon becomes noticeable with the increase of the diffusion coefficient. This illustrates the fact that the diffusion is a highly activated process for the refractive index profile, provided the flow rates of the liquids are the same.

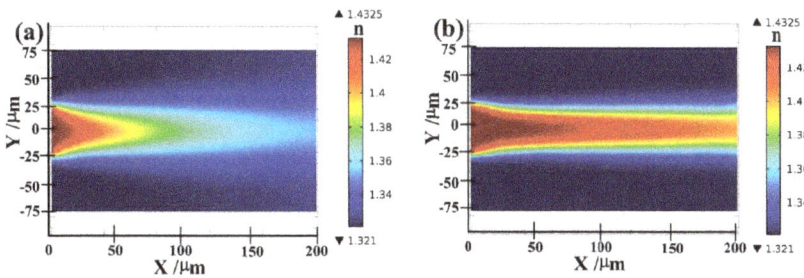

Figure 3. The refractive index profiles in the *xoy* plane with different diffusion coefficient D (with a fixed mass fraction of 0.3, and Q_{clad} = 8, Q_{core} = 8 × 10^3 pL/s). (**a**) $D = 1 \times 10^{-9}$ m^2/s; (**b**) $D = 4 \times 10^{-10}$ m^2/s.

3.2. The Adjustment of the Focal Lengthof the L-GRIN Microlens

It is known that a relatively lower average velocity of liquids offers remarkable diffusion effect, and the focus with an adjustable focal length of the L-GRIN microlens forms by changing the diffusion process. The diffusion process of the liquid is influenced by the average velocity U of liquids and the diffusion coefficient D which is affected simultaneously by the concentration C of the solution. Therefore, the focal length of the L-GRIN microlens can be tuned by adjusting the concentration and the flow rateof the liquids. Hereinafter, the concentration of the ethylene glycol solution is altered through changing the mass fraction of the ethylene glycol solution.

3.2.1. Tuning Mass Fraction of Ethylene Glycol Solution

Because the diffusion coefficient changes with the concentration of the core liquid, the refractive index profile of the L-GRIN microlens varies with the mass fraction of the ethylene glycol solution. As the diffusion progresses, the solution flowing into different regions has a different concentration, and the diffusion coefficient is, therefore, also different. However, according to Equation (4), compared with the concentrated ethylene glycol solution, the diluted ethylene glycol solution will have an obvious gradient refractive index profile whose diffusion coefficient is larger and can be considered approximately constant during the diffusion process. To facilitate the analysis, smaller mass fractions of ethylene glycol ranging from 0.05 to 0.4 were selected and the focal length of the L-GRIN microlens with different mass fractions was calculated while the flow rate of the liquids was kept the same ($Q_{core} = 1 \times 10^3$ pL/s, $Q_{clad} = 8 \times 10^3$ pL/s). Firstly, we simulated the refractive index profiles of the L-GRIN microlens with the mass fraction increasing from 0.05 to 0.4 by 0.05, and simulation results showed that the sharper peak profile of the refractive index could be achieved with the increase of the mass fraction. Figure 4a shows the refractive index profile with the mass fraction of 0.2, and the cross-sectional refractive index profiles at five different locations (x = 50, 100, 150, 200 and 250 μm) are shown in Figure 4b. The beam transmission and focusing process can be simulated by using the ray tracing method based on the refractive index profile, which is shown in Figure 4c. The focal length is the distance from the outlet to the focal point, and Figure 4d shows that the focal length of the L-GRIN microlens can be tuned with different mass fractions of the ethylene glycol solution. When the flow rates of the core and cladding liquid are fixed, the focal length of the L-GRIN microlens varies from 942 to 11 μm with a mass fraction of ethylene glycol ranging from 0.05 to 0.4. Therefore, controlling the mass fraction of the core liquid is a useful way to adjust the focal length of the device.

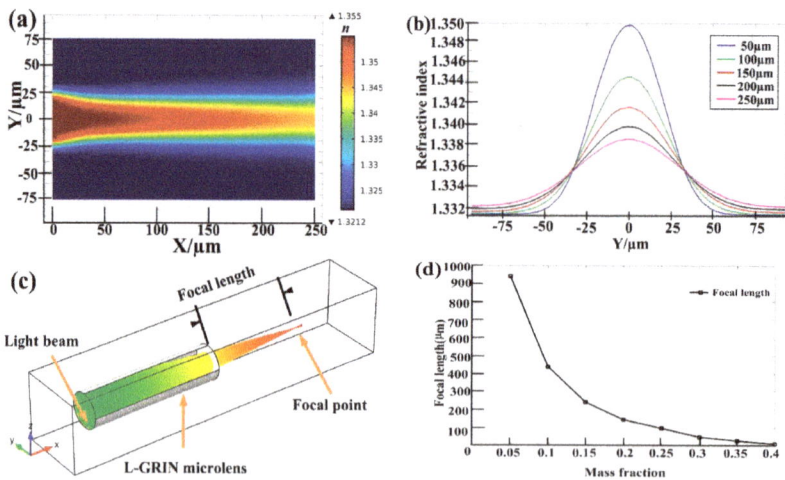

Figure 4. (a) The refractive index profile in the *xoy* plane with the mass fraction of 0.2; (b) The cross-sectional refractive index profiles at five different locations (x = 50, 100, 150, 200 and 250 μm) with $Q_{core} = 1 \times 10^3$ pL/s, $Q_{clad} = 8 \times 10^3$ pL/s; (c) Beam transmission and focusing process; (d) The focal length of the L-GRIN microlens with different mass fractions of the ethylene glycol solution.

3.2.2. Tuning the Flow Rate of Liquids

The refractive index profile of the L-GRIN microlens can also be tuned by changing the flow rate of the liquids. In order to analyze the influence of the flow rate of the liquids on the focal length of the L-GRIN microlens, the ethylene glycol solution is co-injected with DI water into the lens chamber at the same flow rate ranging from 0.5×10^3 to 5×10^3 pL/s. In the analysis, we suppose there is no relative slip between the ethylene glycol solution and the DI water. Additionally, for simplifying the calculation, simulations were started by setting $D = 8 \times 10^{-10} m^2/s$ and $\mu = 1 \times 10^{-3}$ Pa·s with a fixed mass fraction of 0.3.

We simulated the refractive index profiles with the flow rate increasing from 0.5×10^3 to 5×10^3 pL/s by 0.5×10^3 pL/s, and the results demonstrated that the diffusion was no longer dominant with the increase of the flow rate. Figure 5a shows the refractive index profile with the flow rate of 4×10^3 pL/s, and the cross-sectional refractive index profiles at five different locations (x = 50, 100, 150, 200 and 250 µm) are shown in Figure 5b. Figure 5c shows the focal length of the L-GRIN microlens is adjusted with the different flow rate. According to our calculations, the focal length decreases observably from 128 to 8 µm when the flow rate varies from 0.5×10^3 to 5×10^3 pL/s. However, when the flow rate is larger than 5×10^3 pL/s, the converging of the light beam caused by focusing becomes more and more unobvious. Therefore, continuously tuning the flow rate of the liquids in a certain range provides a tunable focal length when the mass fraction of the ethylene glycol is kept constant.

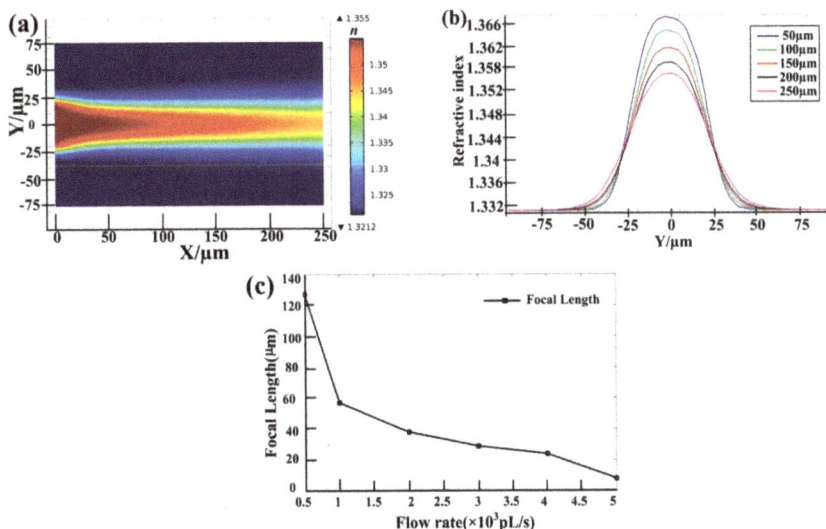

Figure 5. (a) The refractive index profile in the *xoy* plane with the flow rate of 4×10^3 pL/s; (b) The cross-sectional refractive index profiles at five different locations (x = 50, 100, 150, 200 and 250 µm) with $Q_{clad} = 8 \times 10^3$ pL/s, $Q_{core} = 1 \times 10^3$ pL/s; (c) The focal length of the L-GRIN microlens with different flow rates of solutions.

In general, the focal length of the lens is related to the lens radius, the refractive index difference between the lens center and border, and the lens thickness; in addition, it is dependent on the wavelength of the incident light which will change the refractive index profile of the lens. Thus, a shorter wavelength of incident light (when other parameters are constant) results in larger refractive index contrast, which causes light to bend toward the lens axis more significantly and leads to the decreased focal length.

3.3. The Adjustment of the Focal Spot of the L-GRIN Microlens

   The focus becomes a tunable focal spot at a relatively higher average velocity. The adjustment of the focal spot of the L-GRIN microlens can also be achieved by changing the flow rate of the liquids when the average velocity is higher. The case of no relative slipbetween the core and cladding liquids with a lower average velocity has been analyzed above. Analyzing a more sophisticated case will be helpful to better understand the influence of flow rate on the refractive index profile. For instance, the ethylene glycol solution is injected in different flow rates with DI water, andthere is a relative slip between the core and cladding liquids. For analyzing the influence of the relative slip between the core and cladding liquids on the refractive index profile, the cladding flow rate was kept constant (40×10^3 pL/s) and the flow rate of the core inlet varied from 2×10^3 pL/s to 50×10^3 pL/s by 5×10^3 pL/s. Figure 6a shows the refractive index profile with the core inlet flow rate of 25×10^3 pL/s, and the cross-sectional refractive index profiles at five different locations ($x = 50$, 100, 150, 200 and 250 µm) are shown in Figure 6b. It is found that there is little variation of the refractive index in central regions with different cross-sectional lengths along the x axis. The width in the central region along the y axis, in which the refractive index keeps almost constant, is defined as the core width. When the core inlet flow rate is $Q_{core} = 25 \times 10^3$ pL/s, the core width of the refractive index is 24 µm, as shown in Figure 6b. The constant refractive index in the central region hasno obvious effect on the focusing process;thus, the size of the output beam spot can be determined mainly by the core width. The curve with circle in Figure 6c shows the relationship between the core inlet flow rate and the core width. Additionally, the curve width dot shows the changing trend of half the width of the refrative index profile with different Q_{core}. When the core inlet flow rate remainsconstant, with a relatively low core inlet flow rate (lower than 10×10^3 pL/s),the core width approachs zero. In that case, the L-GRIN microlens can theoretically focus the beam onto one point when the ratio of the core inlet flow rate to the cladding inlet flow rate is lower than 0.25. However, the core width increases along with the increaseof the core inlet flow rate (higher than 10×10^3 pL/s); that is, when increasing the ratio of the flow rates of the ethylene glycol solution to the DI water, the size of the output spot may continue to grow. Thus, controlling the ratio of the core inlet flow rate to the cladding inlet flow rate is a useful way to dynamically adjust the size of the light beam in microscale.

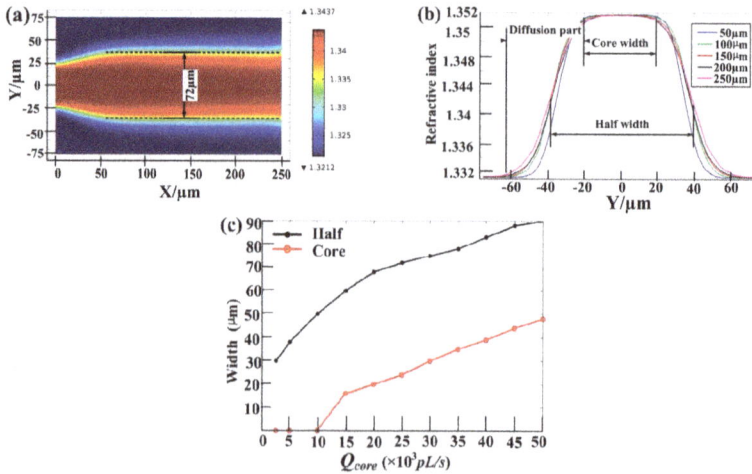

Figure 6. (**a**) The refractive index profile in the *xoy* plane with $Q_{core} = 25 \times 10^3$ pL/s; (**b**) The cross-sectional refractive index profiles at five different locations ($x = 50$, 100, 150, 200 and 250 µm) at $Q_{core} = 25 \times 10^3$ pL/s; (**c**) The variation of core width and half width with different Q_{core}.

Figure 6b reveals that the cross-sectional refractive index profile in the *yoz* plane changes slightly even though *x* is different, which is fitted using six-degree polynomial functions. The optimal parameters of the functions areselected through the test of the partial fitting square sum on the basis of the variance analysis of the fitting equation. For example, the fitted refractive index profile at $x = 125$ µm with $Q_{core} = 2.5 \times 10^4$ pL/s and $Q_{core} = 4 \times 10^4$ pL/s can be expressed as:

$$n = 1.1 \times 10^{-12}s^6 - 2.2 \times 10^{-13}s^5 - 6.45 \times 10^{-9}s^4 - 2.73 \times 10^{-10}s^3 + 3.62 \times 10^{-6}s^2 + 1.99 \times 10^{-6}s + 1.3533 \quad (7)$$

where $s = \sqrt{y^2 + z^2}$ This is a six-order polynomial of *s*, which represents the non-linear relation between the refractive index *n* of the liquids and the chamber position in the *x* direction. Figure 7a shows that the fitting refractive index profile matches well with the simulated data. For analyzing the beam focusing effect of the L-GRIN microlens, the beam transmission and focusing process was simulated using the ray tracing method according to the refractive index profile, which is shown in Figure 7b. A beam spot with the diameter of 23.5 µm and the focal length of 235.3 µm was achieved. The ability of the tunable focusing of the beam is significant for a wide range of applications in integrated optics and lab-on-a-chip systems.

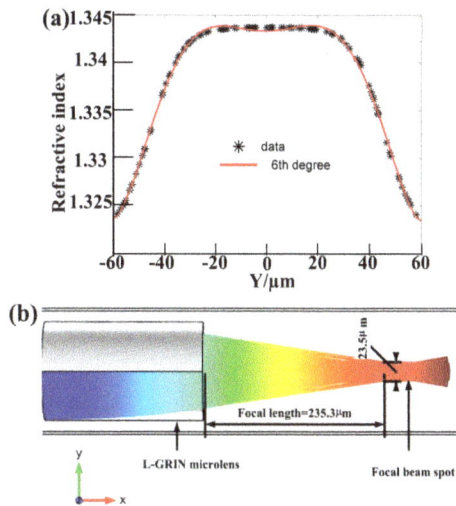

Figure 7. (a) The fitting refractive index profile of the simulated data and six-degree polynomial at $x = 125$ µm with $Q_{core} = 2.5 \times 10^4$ and $Q_{core} = 4 \times 10^4$ pL/s; (b) Beam transmission and focusing process.

In the L-GRIN microlens, the diameter of the focal spot is affected by the focal length and the wavelength of incident light. For a given incident light, one can conclude that the focal spot is the smallest when the focal length is the shortest. As the wavelength increases, the focal length is gradually increased, and the focal spot is increased.

These discussions above are all for a steady-state flow. However, the response time of the focal length and the focal spot change are important parameters for the stability of an adaptive lens operation, which is mainly dependent on the diffusion speed of two inlet flows. According to Equation (1), the response time varies with the concentration *C* and the diffusion coefficient *D* of the liquids, which are also found to be affected by the temperature, the average velocity *U* and the viscosity of the liquids, the structure and materials of the chamber, *etc.* For the fixed inlet liquids and the chamber, as the temperature increases, the response time decreases. However, increasing the average velocity of the liquids would lend to an unfavorable slow response time because the contacting time of the two liquids is shortened. Therefore, for practical applications we need to choose the suitable liquids and optimize the microlens parameters; meanwhile, the microlens material should be carefully chosen.

45

4. Conclusions

We designed an in-plane L-GRIN microlens that can realize two-dimensional light beam focusing dynamically. The diffusion coefficient, mass fraction and flow rate of the core inlet and the cladding inlet are demonstrated to be the main influencing factors for changing the refractive index profile of the L-GRIN microlens, and the influences of these factors on the focusing process of the L-GRIN microlens are studied. It is found that adjusting the mass fraction and flow rate of the ethylene glycol is an efficient way to change the focal length of the output beam. The focal length varies from 942 to 11 μm at the mass fraction of ethylene glycol varying from 0.05 to 0.4. In addition, by varying the core and cladding inlet flow rate from 0.5×10^3 to 5×10^3 pL/s, the focal length of the microlens changes from 127.1 to 8 μm. In addition, the adjustable size of the output beam spot can be provided by controlling the relative slip between the core inlet and the cladding inlet flow rate with a relatively high average velocity. The flexible properties of manipulating the light beam in microscale have advantages for integrated devices in lab-on-a-chip applications.

Acknowledgments: Acknowledgments: This work is supported by the National Natural Science Foundation of China (Grant No. 61172081) and the Natural Science Foundation of Zhejiang Province, China (Grant No. LZ13F010001).

Author Contributions: Author Contributions: Zichun Le and Yunli Sun planned and performed the reported designs and their analysis. Ying Du has written the main manuscript and prepared the figures and tables. All authors reviewed the manuscript.

Conflicts of Interest: Conflicts of Interest: The authors declare no conflict of interest.

References

1. Erickson, D.; Yang, C.; Psaltis, D. Optofluidics emerges from the laboratory. *Photon. Spectra* **2008**, *42*, 74–79.
2. Mao, X.L.; Lin, S.C.S.; Lapsley, M.I.; Shi, J.J.; Juluri, B.K.; Huang, T.J. A tunable optofluidic microlens based on gradient refractive index. *Lab Chip* **2009**, *9*, 2050–2058. [CrossRef] [PubMed]
3. Godin, J.; Lien, V.; Lo, Y.H. Demonstration of two-dimensional fluidic lens for integration into microfluidic flow cytometers. *Appl. Phys. Lett.* **2006**, *89*, 061106. [CrossRef]
4. Wang, Z.; el-Ali, J.; Engelund, M.; Gotsaed, T.; Perch-Nielsen, I.R.; Mogensen, K.B.; Snakenborg, D.; Kutter, J.P.; Wolff, A. Measurements of scattered light on a microchip flow cytometer with integrated polymer based optical elements. *Lab Chip* **2004**, *4*, 372–377. [CrossRef] [PubMed]
5. Martini, J.; Recht, M.I.; Huck, M.; Bern, M.W.; Johnson, N.M.; Kiesel, P. Time encoded multicolour fluorescence detection in a microfluidic flow cytometer. *Lab Chip* **2012**, *12*, 5057–5062. [CrossRef] [PubMed]
6. Yin, D.; Lunt, E.J.; Rudenko, M.I.; Deamer, D.W.; Hawkinsand, A.R.; Schmidt, H. Tailoring the transmission of liquid-core waveguides for wavelength filtering on a chip. *Lab Chip* **2007**, *7*, 1171–1175. [CrossRef] [PubMed]
7. Ozcelik, D.; Phillips, B.S.; Parks, J.W.; Measor, P.; Gulbransen, D.; Hawkins, A.R.; Schmidt, H. Dual-core optofluidic chip for independent particle detection an tunable spectral filtering. *Lab Chip* **2012**, *12*, 3728–3733. [CrossRef] [PubMed]
8. Kuiper, S.; Hendriks, B.H.W. Variable-focus liquid lens for miniature cameras. *Appl. Phys. Lett.* **2004**, *85*, 1128–1130. [CrossRef]
9. Ren, H.W.; Wu, S.T. Variable-focus liquid lens. *Opt. Express* **2007**, *15*, 5931–5936. [CrossRef] [PubMed]
10. Ren, H.W.; Wu, S.T. Tunable-focus liquid microlens array using dielectrophoretic effect. *Opt. Express* **2008**, *16*, 2646–2652. [CrossRef] [PubMed]
11. Ren, H.W.; Xianyu, H.Q.; Xu, S.; Wu, S.T. Adaptive dielectric liquid lens. *Opt. Express* **2008**, *16*, 14954–14960. [CrossRef] [PubMed]
12. Dong, L.; Agarwal, A.K.; Beebe, D.J.; Jiang, H. Adaptive liquid microlenses activated by stimuli-responsive hydrogels. *Nature* **2006**, *442*, 551–554. [CrossRef] [PubMed]
13. Chronis, N.; Liu, G.; Jeong, K.H.; Lee, L. Tunable liquid-filled microlens array integrated with microfluidic network. *Opt. Express* **2003**, *11*, 2370–2378. [CrossRef] [PubMed]
14. Pang, L.; Levy, U.; Campbell, K.; Groisman, A.; Fainman, Y. A set of two orthogonal adaptive cylindrical lenses in a monolith elastomer device. *Opt. Express* **2005**, *13*, 9003–9013. [CrossRef] [PubMed]

15. Tang, S.K.Y.; Stan, C.A.; Whitesides, G.M. Dynamically reconfigurable liquid-core liquid-cladding lens in a microfluidic channel. *Lab Chip* **2008**, *8*, 395–401. [CrossRef] [PubMed]
16. Zickar, M.; Noell, W.; Marxer, C.; de Rooij, N. MEMS compatible micro-GRIN lenses for fiber to chip coupling of light. *Opt. Express* **2006**, *14*, 4237–4249. [CrossRef] [PubMed]
17. Huang, H.; Mao, X.L.; Lin, S.C.S.; Kiraly, B.; Huang, Y.P.; Huang, T.J. Tunable Two-Dimensional Liquid Gradient Refractive Index (L-GRIN) Lens for Variable Light Focusing. *Lab Chip* **2010**, *10*, 2387–2393. [CrossRef] [PubMed]
18. Li, Z.; Zhang, Z.; Scherer, A.; Psaltis, D. Mechanically tunable optofluidic distributed feedback dye laser. *Opt. Express* **2006**, *14*, 10494–10499. [CrossRef] [PubMed]
19. Wolfe, D.B.; Conroy, R.S.; Garstecki, P.; Mayers, B.T.; Fischbach, M.A.; Paul, K.E.; Prentiss, M.; Whitesides, G.M. Dynamic control of liquid-core/liquid-cladding optical waveguides. *Proc. Natl. Acad. Sci. USA* **2004**, *101*, 12434–12438. [CrossRef] [PubMed]
20. Perumal, M.; Raju, R.K.G. Approximate Convection-Diffusion Equations. *J. Hydrol. Eng.* **1999**, *4*, 160–164. [CrossRef]
21. Wu, Z.; Nguyen, N.T. Hydrodynamic focusing in microchannels under consideration of diffusive dispersion: Theories and experiments. *Sens. Actuators B* **2005**, *107*, 965–974. [CrossRef]
22. Kapur, J.N.; Shukla, J.B. Flow of incompressible immiscible fluids between two plates. *Appl. Sci. Res. A* **1964**, *13*, 55–60. [CrossRef]
23. Yang, Y.; Liu, A.Q.; Chin, L.K.; Zhang, X.M.; Tsai, D.P.; Lin, C.L.; Lu, C.; Wang, G.P.; Zheludev, N.I. Optofluidic waveguide as a transformation optics device for lightwave bending and manipulation. *Nat. Commun.* **2012**, *3*, 651. [CrossRef] [PubMed]
24. Ternstrom, G.; Sjostrand, A.; Aly, G.; Jernqvist, A.J. Mutual diffusion coefficients of water + ethylene glycol and water + glycerol mixtures. *Chem. Eng. Data* **1996**, *41*, 876–879. [CrossRef]
25. Wang, M.H. Measurement of Binary Mutual Diffusion Coefficient of Several Aqueous Glycol Solutions. Master's Thesis, Chung Yuan Christian University, Taiwan, July 2009. (In Chinese)

micromachines

MDPI

Article

Fiber-Based, Injection-Molded Optofluidic Systems: Improvements in Assembly and Applications

Marco Matteucci [1], Marco Triches [2,3], Giovanni Nava [4], Anders Kristensen [1], Mark R. Pollard [2,*],
Kirstine Berg-Sørensen [5,*] and Rafael J. Taboryski [1,*]

[1]   Department of Micro- and Nanotechnology, Technical University of Denmark, Ørsteds Plads, Building 345B,
Kgs. Lyngby 2800, Denmark; mamat@nanotech.dtu.dk (M.M.); anders.kristensen@nanotech.dtu.dk (A.K.)
[2]   DFM A/S Matematiktorvet 307, Kgs. Lyngby 2800, Denmark
[3]   Department of Photonics Engineering, Technical University of Denmark, Ørsteds Plads, Building 343,
Kgs. Lyngby 2800, Denmark; matri@fotonik.dtu.dk
[4]   Department of Medical Biotechnology and Translational Medicine, Università degli Studi di Milano,
Milan 20122, Italy; giovanni.nava@unimi.it
[5]   Department of Physics, Technical University of Denmark, Fysikvej, Building 309,
Kgs. Lyngby 2800, Denmark
*   Correspondence: mp@dfm.dk (M.R.P.); kirstine.berg-sorensen@fysik.dtu.dk (K.B.-S.); rata@nanotech.dtu.dk
(R.J.T.); Tel.: +45-4525-5823 (M.R.P.); +45-4525-3101 (K.B.-S.); +45-4525-8155 (R.J.T.)

Academic Editors: Shih-Kang Fan, Da-Jeng Yao and Yi-Chung Tung
Received: 7 November 2015; Accepted: 4 December 2015; Published: 9 December 2015

Abstract: We present a method to fabricate polymer optofluidic systems by means of injection
molding that allow the insertion of standard optical fibers. The chip fabrication and assembly
methods produce large numbers of robust optofluidic systems that can be easily assembled and
disposed of, yet allow precise optical alignment and improve delivery of optical power. Using a
multi-level chip fabrication process, complex channel designs with extremely vertical sidewalls, and
dimensions that range from few tens of nanometers to hundreds of microns can be obtained. The
technology has been used to align optical fibers in a quick and precise manner, with a lateral alignment
accuracy of 2.7 ± 1.8 µm. We report the production, assembly methods, and the characterization
of the resulting injection-molded chips for Lab-on-Chip (LoC) applications. We demonstrate the
versatility of this technology by carrying out two types of experiments that benefit from the improved
optical system: optical stretching of red blood cells (RBCs) and Raman spectroscopy of a solution
loaded into a hollow core fiber. The advantages offered by the presented technology are intended to
encourage the use of LoC technology for commercialization and educational purposes.

Keywords: fiber-based optofluidics; injection molding; optical trapping; hollow core fiber enhanced
Raman spectroscopy

1. Introduction

Since 2005 [1] merging of optics and fluidics at the micro and nanoscale for Lab-on-Chip (LoC)
purposes opened a wide range of opportunities both in basic and applied research [2]. Despite many
demonstrations of technical feasibility and the continued broadening of the field, few devices with
optical functionalities have reached the market. The limited commercialization of this technology lies
in the fact that, at present, the majority of LoC optofluidic systems are fabricated with techniques
that are not production-friendly either in terms of materials or in terms of optical elements: Existing
microfluidic systems in glass are characterized by high production cost and, in addition, optofluidic
systems with waveguides produced using femtosecond laser machining [3] or DUV writing [4] suffer
from high optical losses. Alternatively, low-loss, optical fiber-based systems made in soft polymer

materials like polydimethylsiloxane (PDMS) require more laborious procedures such as pneumatically driven active optical fiber manipulators [5] for optical alignment on-chip because of the fiber housing deformation. The ideal system would be easy to align with high precision, mechanically strong, easy to interface with both fluidics and optics, have negligible biofouling, deliver high power *in situ* and would be disposable.

To make a substantial improvement towards such systems, we developed injection molded (*i.e.*, production-friendly) optofluidic chips in hard Cyclic Olefin Copolymer (CoC) TOPAS 5013 with embedded commercially available optical fibers (Figure 1a,b). We consider TOPAS 5013 to be an ideal polymer for fabrication of LoC systems because of its high transparency in the visible wavelength range [6], its high glass transition temperature (140 °C), its low water absorption and its resistance to acids and alkaline agents, as well as to polar solvents [7]. TOPAS also avoids undesired biofouling with minimal surface treatment requirements [8] and autofluorescence of TOPAS 5013 can be reduced to values that are about 20% lower than the ones of silica chips [9].

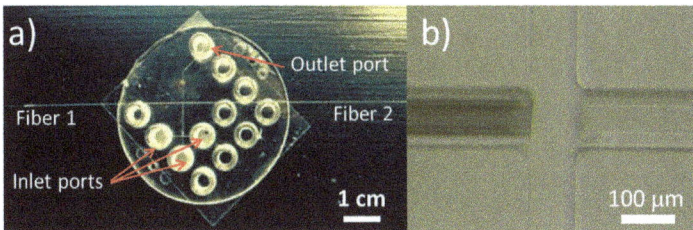

Figure 1. (**a**) Optofluidic chip in TOPAS 5013 L10 polymer with a system for hydrodynamic focusing of cells and embedded optical fibers. Optimal coupling with external fluidics is ensured by the presence of on-chip standard Luer connectors. The chip was designed with three inlet ports to allow hydrodynamic focusing of cells. (**b**) Detail of chip with a photonic crystal fiber (Left) and single-mode fiber (Right).

Compared to earlier demonstrations of optofluidic devices with embedded commercial optical fibers [10,11] the reported chip fabrication process is production ready and allows the height of the optical beam path relative to the microfluidic channel to be tuned easily in the design phase. Moreover, novel process modifications involving a double positive resist exposure (described in the "Materials and Methods", Section 4.1) allowed the fabrication of the novel auto-aligning two-level square geometry for fiber housing. For our purposes, we embedded standard bare optical fibers with a nominal diameter of 125 μm and interfaced them with a microfluidic channel of 100 μm × 100 μm section (Figure 1b), thus producing an optical path at a height of around 40 μm above the channel surface. The dimensions of the fluidic channel were chosen to balance the requirements of hydrodynamic resistance and of delivery of the optical power to the center of the channel (for optical stretching). The chosen distances also proved to be a good compromise for delivery of power across the channel (for the Raman sensing). Uniquely, the fiber grooves have a square design ensuring reproducible alignment by constraining the lateral fiber movement. In general, any fiber with a diameter of 125 μm could be inserted in the chip described here, whereas other dimensions require a simple modification in the design phase. To validate the chip functionality and versatility, we present results obtained through the utilization of the chips for both optical stretching of red blood cells (RBCs) and for in-fiber Raman spectroscopy of liquids. Both techniques require a high level of alignment precision and coupling [12], for this reason they are considered ideal examples for the application and validation of the technology.

2. Results

2.1. Fabrication Results

Figure 2 shows a scanning electron microscopy image (Figure 2a) of the Ni stamp (here referred to as shim) together with optical (Figure 2b) and mechanical (Figure 2c) profilometry data of the polymer chip. A groove width of 128 µm was measured by means of mechanical profilometry with an uncertainty of 2.5 µm, which is compatible with the designed width of the fiber groove (125 µm). In the final samples, while the difference in height between the fiber groove and the channel was measured to be around 25 µm (Figure 2b), the total depth of the groove of the final systems was around 135 µm (Figure 2b,c).

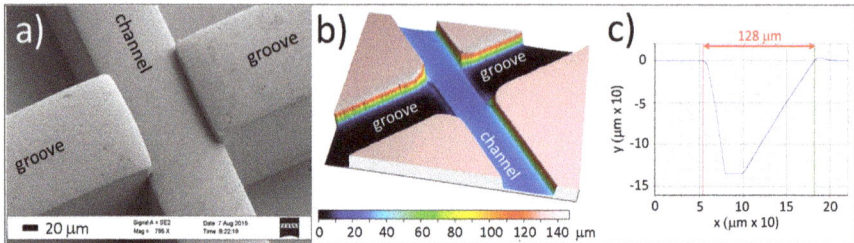

Figure 2. (**a**) SEM micrograph of the Ni shim in the area where the fiber grooves meet the channel; (**b**) Confocal microscope profilometry of the polymer chip that shows the channel depths and the difference between the groove and the channel; (**c**) Mechanical profilometry of the fiber grooves inside a polymer chip. NB, this is a convolution of groove dimensions and dimensions of the profilometer tip (5 µm in diameter).

In order to properly seal the fibers in the channels, a higher bonding temperature (130 °C instead of 125 °C) was chosen to allow controlled swelling of the polymer lid inside the fiber groove during bonding and the formation of a tight microfluidic seal. The effect of bonding the lid onto the optofluidic chip was observed by the cross-section of an assembled chip (which was assembled following the method reported in Section 4.1). In this case, the lid was a 2 mm thick 2-inch of TOPAS wafer and it was thermally bonded to the chip, which permitted cutting and polishing of the test sample. The assembled chip was cut in half using a jeweller's saw, polished using diamond lapping paper (Thorlabs LF6D, Newton, NJ, USA) and finally observed using a microscope (Nikon, ×40 objective lens, Tokyo, Japan) as shown in Figure 3. Green laser light was coupled into one end of a single mode fiber and was observed at the other fiber end in the chip cross-section (see Section 2.3 for further details on the coupling of laser light into the fiber).

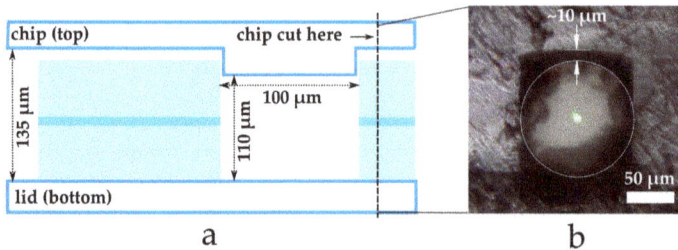

Figure 3. Effect of lid bonding on assembled chip: (**a**) Design of an assembled chip; (**b**) cross-section of assembled chip including single mode fiber fixed within the fiber groove, transmitting green laser light. The outline of the fiber has been enhanced to make it visible.

The image shows that the width (*i.e.*, horizontal dimension in Figure 3b) of the groove fits well to the diameter of the fiber (125 μm), whereas the depth of the groove is approximately 10 μm greater than the fiber diameter. This gap was deliberately included in the chip design to accommodate swelling of the lid during its bonding to the chip, but can act as a cause of fiber misalignment.

Ten chips were assembled and the distance between the ends of the two cleaved fibers was measured for all chips using a microscope (Nikon, ×20 objective lens). The average distance between the ends of the two cleaved fibers is 104.6 with a maximum uncertainty of ±2.4 μm.

We tested the fiber alignment by checking the optical power transmitted through one unfilled chip (*i.e.*, no fluid in between the fibers). The chip was assembled using two single-mode fibers (Thorlabs SM980), equipped with standard fiber connectors. A fiber laser emitting at 1550 nm wavelength was used to perform the test. The injected power was measured to be 15.19 mW, while the output was 8.03 mW. Using this information and neglecting the fiber loss (less than 0.001 dB/m at this wavelength as reported in the datasheet), we calculate the transmission across the empty junction to be 53%. Comparing this result with the theoretical loss expected in a fiber-to-fiber coupling (as described in Materials and Methods Section 4.3), we retrieve a lateral misalignment of 2.7 ± 1.8 μm. As shown in Figure 3b, we believe this fiber misalignment most likely occurs in a vertical direction.

2.2. Single Chip Applied in Optical Stretching

Optical stretching [13] is highly relevant for the study of mechanical properties of single cells. Mechanical properties of cells have been shown to be closely related to the health of the studied biomaterial [14]. Here, for demonstration, we trapped and stretched single red blood cells (RBCs). The embedded single-mode (SM) fibers were connected to two independent diode lasers, we prefer the independent lasers as we may therefore easily tune the optical forces from each by varying the power of the two lasers. We took care to use two identical lasers to ensure similar time response. Both the laser powers and pump flow rates could be controlled using a custom made LabView code. To see the trapping region and recording the images of the trapped objects, the chip was mounted on an inverted microscope with a CCD camera. The RBCs were suspended in a hypotonic buffer solution with low osmotic pressure in order to obtain spherical RBCs of 8 μm diameter. The minimum power to trap the cells was found to be around 100 mW from each laser diode.

Once a cell was trapped, the laser power was increased in four to seven steps up to the maximum value (450 mW output from the laser), while returning to the minimum power between each step to allow the cell to relax. For each laser power, the images of the cell were recorded and both main and minor axes were measured. The ratios of the major and minor cell diameters were then determined using an image processing code and displayed as a function of the total laser power from two lasers (Figure 4). The data obtained reveals that the ratio between main and minor axes of RBC changes linearly with applied power in the power range used. As expected, the axial scattering forces, in the direction of propagation of the counter propagating beams, act to provide a stable trapping point in the axial direction and to stretch the cells (direction of R_1) whereas the transversal force, due to the field gradient perpendicular to the optical axis (direction of R_2), merely assists to capture and stably trap the cells in the transversal direction [15]. The results in Figure 4 are in good agreement with results obtained by Guck et al [16]. The simplicity of use was verified during the PolyNano summer school 2014 [17] where students with no previous experience could easily assemble the described chips and use them for cell stretching experiments with only two days of work. Four student groups produced four different chips, all with well aligned fibers, and trapped polystyrene beads and/or red blood cells in these chips.

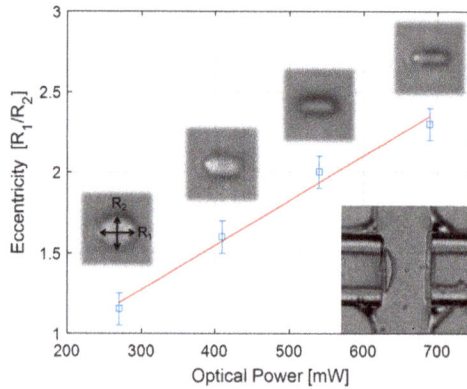

Figure 4. Stretching of a single RBC. The stretching ratios, R1/R2 of a RBC in the optical stretcher as a function of total stretching laser power. The major and minor axes of the cell were measured using the corresponding images shown as insets. Resulting values and error bars are the result of edge-detection in 30 images recorded at the same laser power. In the lower right corner, the inset shows a black and white image of a red blood cell trapped between the two SM optical fibers embedded in the microfluidic chip.

2.3. Dual Chip Applied in Fiber-Based Raman Spectroscopy

Our polymer chip technology was used to create a new experimental setup for the Raman spectroscopy of liquids. Our system was made by the quick and simple integration of a hollow core fiber (HC fiber) [18] between two chips, where the hollow core can be filled with different liquid samples by pumping liquid from alternate ends of the HC fiber, as described by Khetani *et al.* [8]. The system (including the HC fiber) was initially filled by pumping milli-Q water into Chip 1 at a rate of 230 μL/min (see Materials and Methods for experiment diagram). A transmission of 10% was achieved, and the Raman spectrum for the sample (Figure 5, black curve) showed the expected peaks created by Raman scattering from the water molecules loaded in the HC-fiber and the silica molecules present in all the optical fibers.

A transmission of 10% was achieved, and the Raman spectrum for the sample (Figure 5, black curve) showed the expected peaks created by Raman scattering from the water molecules loaded in the HC-fiber and the silica molecules present in all the optical fibers.

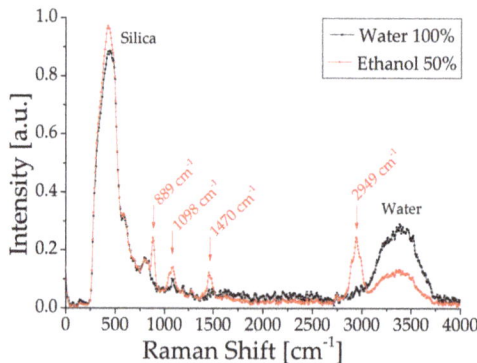

Figure 5. Raman spectra for solutions filled into hollow core fiber, showing ethanol solution (**red**) and milli-Q water (**black**).

Next, the HC fiber was refilled with an ethanol solution (50:50 water/ethanol by volume) which was flowed into Chip 2 using the same flow rate. The filling process was monitored by observing changes in the spectra, after 90 min the spectra did not vary further, indicating that filling was complete. This filling time given the calculated pressure difference across the HC fiber (17.2 kPa) agrees with previous work [8]. The pressure difference across the two chips is calculated based on the dimensions of the chip's microfluidic channel and the microstructure of the HC-fiber. The viscosity of the ethanol solution was calculated using the Refutas equations [19]. The low pressure applied across the fiber was chosen to characterize the filling time. The overall filling time can be reduced by increasing the pressure thanks to the mechanical properties of the optofluidic system fabricated (more details in the Discussion, Section 3).

The resulting Raman spectrum contained additional signals from the ethanol solution (Figure 5, red curve) that agreed with reported spectra [20,21], and a reduction in the water peak due to dilution with ethanol was observed. All spectra are based on an average of five spectra recorded with an integration time of 2 s, with a three-point moving average applied to reduce spectrometer noise. For comparison purposes, the spectra were normalized using 30 data points that occurred across the broad Silica peak at 500 cm^{-1}.

3. Discussion

The technology proposed demonstrates its versatility thanks to the two different optofluidic applications presented. The chip technology has such high repeatability that multiple chips can be concatenated together (as demonstrated in the Raman experiment). In order to balance between the requirements of hydrodynamic resistance and of delivery of the optical power, overlap of deep channels and removal of solid wall separation are required. This marks a difference with previous work [22,23]: in order to precisely overlap deep channels with slight difference in depth, a modified multilayer process is required. The reason is that once a deep channel is dry etched into silicon, the spin coating of a uniform layer of resist for an aligned UV exposure is not possible due to the resist entering the deep channel.

The square shape of the fiber grooves was chosen to reduce the volume surrounding the fiber (with respect to a V-shape), reducing possible leaking and improving the fiber gluing. Grooves with very sharp edges are obtained thanks to the extreme verticality of the DRIE process used. Additionally, the square shape combined with the asymmetric dimensions of the groove (125 μm width × 135 μm depth) allows to compensate the effect of the bonding onto the lid. The lid surface is expected to swell inside the channel/groove due to the force/temperature combination applied in the bonding procedure. The effect has been quantified on the order of 6–8 μm in previously unpublished work, using 100 μm thick lid. Although the bonding procedure developed works successfully with 100 μm and 2 mm thick lid, the effect may vary with the lid thickness. As reported in Figure 3, the 2 mm lid does not seem to suffer of this effect at the higher temperature applied.

All the chips tested were assembled by gluing the fiber in place and subsequently bonding the lid, but the chip design also allows the lid to bond in the first step and subsequently inject the fiber and glue it in place. The first procedure is easier to implement because it does not require special skills on the fiber handling, making it preferable for the didactic purposes of the chip design. The average distance between the ends of the two cleaved can be influenced by mechanical traction applied on the fiber when the glue is not completely cured and can be detrimental for the applications, because of the poor optical coupling in the fiber-to-fiber junction.

The second approach can solve this issue but the fiber insertion requires extreme caution to avoid possible damage of the fiber tip, which will compromise the optical properties of the system as well. For this reason, a novel mask design with a larger section of the groove at the fiber entrance is required. Additionally, the depth of the groove needs to be optimized to minimize the causes of misalignment. Bonding was performed with lids ranging in thicknesses from 100 μm to 2 mm before the insertion and gluing of the fibers. Not only did the placement of the fibers after chip sealing provide a simplified alternative assembly but the bonding of thicker lids makes it possible to have systems with resistance to higher pressures (up to 9 bar with a 100 μm lid) [23]. Thanks to this feature, the filling time of the HC fiber can be reduced down to few minutes, which will be important for the feasibility of a sensor development.

The nominal fiber alignment value (2.7 μm) is due to the vertical shift in the fiber groove upon lid bonding and the considerable uncertainty associated to this value (±1.8 μm) is due to the non-Gaussian distribution of the electromagnetic fields. This fact contributes to the variation of the effective optical properties of the fiber (e.g., the mode field diameter), which affects the calculations. This is especially true for hollow core fiber where the variation is greater than standard optical fiber [24]. We would like to investigate this variation further in the future.

Additional inlet ports were included in the chip design with the intention to produce a sheath flow, and thus hydrodynamic focusing of the red blood cells was performed towards the center of the channel. The presence of additional Luer ports also helped to remove the air from the chip during the Raman test, reducing the risk of bubble formation.

3.1. Technology Benefits for Optical Stretching

The injection molded chips offer clear advantages for the optical stretching application. For trapping and stretching of stiffer cells than the red blood cells, e.g., cancer cells, embedded low-loss fibers with ensured alignment in a disposable chip provides efficient stretching even with inexpensive single mode fiber coupled diode lasers. The disposability reduces the risk of any cross-contamination between samples that would otherwise, in a more costly microfluidic system, be investigated in the same chip. In a given optical stretching experiment, the surface stress applied to the trapped cell depends on the distance between the two fibers in a non-trivial fashion, and a new calculation of the surface stress is required for each new distance. The high reproducibility in fiber-to-fiber distance reduces this task.

3.2. Technology Benefits for Raman Spectroscopy

The technology reported here represents a new miniaturized system for the spectroscopic measurement of liquids. We chose Raman spectroscopy due to its known capabilities for liquid identification and measurement of analyte concentration [8]. Previous miniaturized Raman systems based on PDMS microfluidic chips and embedded optical fibers [25] required hundreds of milliwatts of input laser power. In contrast, our system uses laser powers that were a factor of 10 lower. This improvement is because of the precise alignment of the liquid sample and the laser light, and the increased interaction between the laser and the liquid confined in the hollow core fiber (HC-fiber).

The signal enhancement from liquids loaded into a HC fiber with liquid has been widely reported [8,26]. Although these results are remarkable, real-world application of the technology (e.g., in a production-line) requires a simpler, cheaper system that has stable and compact optical components that do not require frequent, time-consuming alignment to light sources and detectors. Our system solves these problems, as it can be easily coupled to fiber-based equipment and it provides a way to encapsulate and fill a HC fiber with liquid that is both low-cost and production ready. Additionally, a smaller gap between the fiber facets will be considered in order to enhance the optical transmission for the Raman applications.

4. Materials and Methods

4.1. Chip Fabrication and Assembly

The multilayer stamp (shim) for injection molding was fabricated with standard cleanroom techniques [22,23,27], these references also describe further details of the shim fabrication and polymer injection molding process. The mask-design is available in the Supplementary Information. This fabrication process allows a precision of replication of the order of 5 nm for 100 nm deep channels [22]. A 100 nm oxide layer is thermally grown on a Si wafer. To enhance the resist adhesion, a hexamethyldisilazane (HMDS) coating is followed by the deposition of a 10 μm thick AZ resist layer. A first UV lithography (Figure 6a) and wet etching of the oxide are performed to pattern the fiber grooves. Multiple masking is achieved by performing a second aligned exposure to pattern the fluidics on the same AZ resist layer and by leaving the oxide film untouched (Figure 6b).

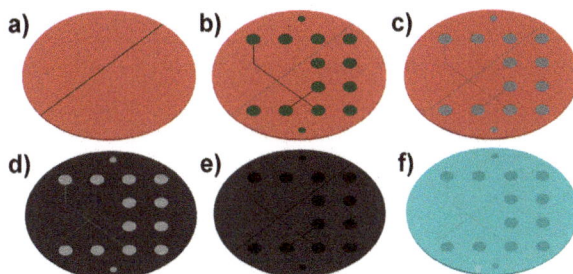

Figure 6. Fabrication process of the microfluidic polymer chip: (**a**) Si wafer oxidation and UV lithography for the opening of the groove for the embedding of optical fibers; (**b**) Wet etching of the oxide layer in yellow-light environment and aligned lithography (on the same resist previously used) for the patterning of the microfluidics; (**c**) Deep reactive ion etching (DRIE) of the groove for optical fibers, dry etching of the oxide in the microfluidic area and DRIE of both the microfluidic channel and the groove; (**d**) Resist stripping and NiV sputterning; (**e**) Ni electroplating and Si wafer removal in KOH; (**f**) Injection molding of CoC polymer.

This is made possible by keeping the sample under yellow light throughout the first two UV lithography steps. A 25 μm deep reactive ion etching (DRIE) of the waveguide grooves is performed while the microfluidic pattern is masked by the oxide.

Removal of the oxide by dry etching is followed by DRIE etching of 110 μm depth of both waveguide grooves and channel (Figure 6c). The width of the fiber grooves was designed to host fibers with a nominal external diameter of 125 μm and to minimize their lateral displacement. The difference in height between groove and channel described in Figure 3 is designed to stop the fibers at a fixed point along the optical axis thus giving a reproducible fiber separation.

Sputtering (Figure 6d) of an adhesion layer of a nickel-vanadium alloy (NiV, 7% vanadium) was then performed. This alloy was chosen because of its much lower magnetization in comparison to pure Ni [28]. This characteristic makes it more suitable for sputtering than pure magnetic Ni [29]. After Ni electroplating and Si etching in KOH the final shim is obtained (Figure 6e). The standard thickness for the shims is between 320 and 340 μm. Although in-depth studies of shim wear were not performed so far, shims with channel sections of 100 nm × 100 nm were used to produce samples in numbers of a few thousands without any functional failure [22]. Final injection molding of TOPAS 5013L10 CoC polymer chips (Figure 6f) is performed with a cycle time of 1 min, suitable for production purposes.

After chip fabrication, fibers were prepared, inserted into the injection molded chip and glued in place using super glue. The chip was then sealed either by thermal bonding of a TOPAS 5013 foil [23] or by gluing of a commercially available poleolefin foil (900320 by HJ-Bioanalytik, Erkelenz, Germany). While the poleolefin foil was found to be enough for the low pressures required by the optical stretching experiments, a tighter seal was required to inject liquid inside the hollow core fibers. Thermally-bonded chips of TOPAS 5013 were able to withstand up to 900 kPa of pressure [23], for this reason the thermal bonding was preferred to the poleolefin foil gluing in the chips dedicated to Raman spectroscopy measurements. The complete process (waveguide insertion and bonding) lasts typically between 60 and 90 min including fiber preparation (5–10 min), insertion and gluing of the fibers (20–30 min), glue curing (30 min), and thermal bonding of the lid (10–20 min).

4.2. Optical Alignment

The optical alignment of the two fibers can be determined from measuring optical power transmitted across one chip. The optical power loss that occurs between two fibers can be calculated from the degree of mismatch between electromagnetic (EM) fields in the optical fibers. For our calculation, the major contribution to this mismatch occurs as the laser beam expands when propagating in free-space across the microfluidic channel. The calculation presented here refers to the measurement reported in Section 2 where an empty chip was used. The radius of the laser beam ω_1, after propagating a distance z in free-space is calculated using the formula:

$$\omega_1 = \omega_0 \sqrt{1 + \left(z\lambda/\pi\omega_0^2\right)^2} = 11.2 \ \mu m \tag{1}$$

where $\lambda = 1550$ nm and $z = 104.6$ μm as reported in Section 2.1. Assuming that the EM field has a Gaussian profile, the theoretical coupling efficiency across the empty channel is given by [30].

$$\eta_{th} = \frac{4\omega_1^2\omega_0^2}{\left(\omega_1^2 + \omega_0^2\right)^2} e^{-2x^2/\left(\omega_1^2 + \omega_0^2\right)} \tag{2}$$

where $\omega_0 = 5.2$ μm is the beam radius of the single mode fiber at 1550 nm and x is the lateral misalignment between the two fiber cores. By using power measurements reported in Section 2.1, we can calculate the lateral misalignment, setting the value of the measured coupling (or transmission) coefficient $\eta_m = 0.53$, and imposing $\eta_m = \eta_{th}$:

$$x = \sqrt{-\frac{\left(\omega_1^2 + \omega_0^2\right)}{2} \ln\left(\frac{\left(\omega_1^2 + \omega_0^2\right)^2}{4\omega_1^2\omega_0^2}\eta_m\right)} \tag{3}$$

Using Equation (3) we obtained a maximum lateral misalignment of 2.7 ± 1.8 μm. The reported maximum error is calculated using the propagation of the uncertainties. The modes field radius used in this calculation (and its relative uncertainty) is taken from the fiber datasheet. The major contribution to the error in this calculation is given by the uncertainties of the mode field radius reported in the datasheet $\omega_0 = 5.2 \pm 0.4$ μm in relation to the effective mode field radius delivered by the fiber, resulting in a corresponding change in the results obtained. Similar results can be obtained using a chip filled with water but the different refractive index and the optical loss of the medium (*i.e.*, water) needs to be taken into account, further reducing the accuracy of the calculation proposed here.

The presence of liquid in the channel between the two fibers has the effect of reducing the laser beam divergence as it travels across the liquid-filled channel and ω_1 is smaller as a result, therefore lowering the loss associated with the EM field mismatch. However, there is unwanted absorption of the laser light by the liquid, a typical value for this loss is <1% (in relation to the input power) for light with a wavelength of 1064 nm travelling over the channel distance of 100 μm, and this loss is lower still for a wavelength of 532 nm. Importantly, both these losses are small when compared to the loss due to lateral fiber misalignment. Here the coupling coefficient reported earlier ($\eta_m = 0.53$) can be expressed as a loss of 47% (in relation to the input power).

4.3. Optical Stretching Setup

In the optical trapping setup (Figure 7), the embedded fibers (Thorlabs SM980, Newton, NJ, USA) were spliced with single mode patch cables with FC/APC connectors (P3-980A-FC/APC) and connected to two independent fiber-coupled diode lasers (Lumics, Berlin, Germany, LU1064M450 with 450 mW maximum power and 1064 nm wavelength) using standard connectors. The laser diodes were controlled by a custom-built power supply and laser control system based on commercial controllers (Thorlabs ITC110) and connected to the computer via LabJack U-3 units. The flow was controlled by a four-channel Fluigent micropump (MFCS-EZ, 0–345 mbar, Villejuif, France). Sample vials, 2 mL centrifuge tubes (Fisher Scientific, Roskilde, Denmark), were filled with physiological salt water (NaCl 9 mg/mL, Fresenius Kabi AG, Bad Homburg, Germany) and DI water in a 50:50 mixture for the two side-inlets, and red blood cells, diluted 1:1000 in the same 50:50 mixture of physiological salt water and DI water, for the middle inlet. The red blood cells were obtained from a same-day sample of fresh blood from an anonymous healthy donor. These vials were mounted in a four-channel Fluiwell holder and connected to the relevant Luer fittings of the chip using standard polymer tubing. The microscope (Leica DMI3000B, Wetzlar, Germany) was equipped with a CCD camera (Thorlabs DCC1240M) mounted on the side-port and cells were visualized with a 40× objective (Leica HCX PL FL L 40×/0.60 CORR PH2 0–2/C). Both the currents to drive the two laser-diodes, the Fluigent flow control system, and the CCD camera were controlled using a custom made LabView code. Image analysis was conducted either with a custom-written MatLab code or in ImageJ.

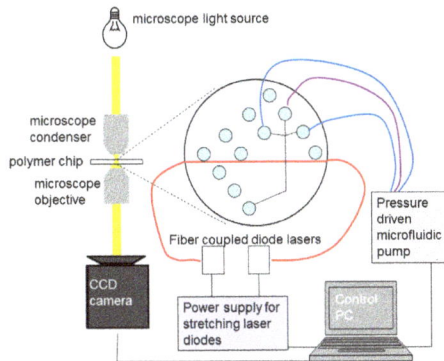

Figure 7. Schematic representation of the optical stretching setup.

4.4. Raman Experiment Setup

The scheme of the setup for Raman experiments is shown in Figure 8. The optical system consisted of a green laser (Coherent Verdi, Santa Clara, CA, USA, 532 nm wavelength) which was coupled using an objective lens (Olympus RMS10X-PF, Tokyo, Japan, ×10 magnification, NA 0.3) and a 3-axis (XYZ) translation stage (Thorlabs MBT616D) to an input fiber (single-mode, SMF-28) which was inserted in the first chip. The maximum laser power launched into our system was 30 mW. An output fiber (multi-mode, Thorlabs FG050LGA) was inserted into a second chip and a HC fiber (NKT Photonics HC-1060-02, Birkerød, Denmark, 10 μm core diameter, 15 cm long) bridged the two chips. The HC fiber was chosen because of its favorable transmission properties after filling with water [31].

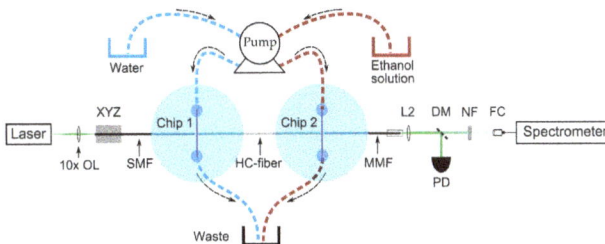

Figure 8. Diagram of optical and fluid system used in the Raman experiments. 10× OL: 10× objective lens. XYZ: 3-axis stage for fiber alignment. SMF: single-mode fiber. MMF: multi-mode fiber. L2: lens. DM: dichroic mirror. NF: notch filter. PD: photodetector used to monitor the power. FC: fiber coupler. Dashed-arrows represent the flow direction.

At the output of the system, a dichroic mirror (Semrock Di02-R532-25×36, Rochester, NY, USA) and a notch filter (Thorlabs NF533-17) were used to separate the Raman scattered light from the pump laser light, which was coupled into a compact, fiber-coupled spectrometer (Thorlabs CCS200). The light reflected by the dichroic mirror was used to monitor the power output over time with a photodetector. The liquid filling system consisted of a peristaltic pump (Peristar Pro, WPI, Sarasota, FL, USA) and the chips were connected using standard connections (Male Luer lock with 1/16″ barb fitting) and tubing (Tygon R3603, Saint Gobain Performance Plastics, Charny, France). Referring to Figure 1a), only the central inlet port was used for the filling. To allow air to escape from the chip during the initial filling, the other two inlet ports were open. During the experiment, these inlet ports were sealed with Luer caps.

5. Conclusions

A novel fabrication scheme for an optofluidic system with standard optical fibers that can be fabricated in large numbers has been reported. The system is intended to be fully fiber-based, removing the need for cumbersome alignment procedures and allowing external connection to fiber-coupled equipment (e.g., laser source, spectrometer, *etc.*). This produces a user-friendly optofluidic device that is quick and easy to assemble with highly reproducible optical characteristics.

Because of this high reliability, we have quickly obtained experimental setups for optical stretching, and by concatenating multiple chips, we created a system for in-fiber Raman spectroscopy. Therefore, we propose this technology as a practical solution for the training of students in LoC techniques.

We demonstrated a lateral fiber alignment to within 3 µm, where the main cause of misalignment is introduced by unwanted vertical movements in the fiber position during lid bonding. We suggest two improvements can be made in the design and manufacture of the chip: reducing the depth of the fiber groove to 125 µm and better control of lid swelling by optimizing the lid bonding recipe. Two further improvements in assembly can also be made: insertion of the fibers after lid bonding and the use of mechanical supports when inserting fibers.

Our fiber-based optofludic system allows high optical powers to be delivered to specimens without damaging the LoC system, thereby broadening the range of possible experiments (including stretching).

Further integration of advanced fiber-based components, e.g., fiber-based Bragg gratings [32], with the existing optical components, would allow the separation of signals of interest (e.g., Raman) and minimizing unwanted background signals (e.g., from silica). Further refinement of chip design (e.g., optimization of distance between the ends of the two cleaved fibers) will enhance the detection range and sensitivity for in-fiber Raman spectroscopy and other spectroscopic techniques. The technology and concepts reported here could be further developed into a compact system for in-line process measurements, for example measuring organic solvents in beverages [20] and protein concentration in media [6], where small volumes can be repeatedly assessed after appropriate flushing/cleaning methods. Fiber-based optofludics allow high optical powers to be delivered to specimens without damaging the LoC system, thereby broadening the range of possible experiments (including stretching).

Finally, the multilevel fabrication scheme described here can also be used to implement more complex geometries where optical and fluidic elements are separated [10,11], thus significantly broadening the number of possible LoC applications of the technology including optogenetic stimulation and detection [32], shape recognition [3], and flow cytometry [33].

Supplementary Materials: The following are available online at http://www.mdpi.com/2072-666X/6/12/1468/s1, Mask CAD design: OpTwe_01.cif, Video of cell stretching: Cellstr_rbc.wmv.

Acknowledgments: This work was supported by the Danish Council for Strategic Research through the Strategic Research Center PolyNano (grant no. 10-092322/DSF). The authors would like to acknowledge the PolyNano summer school, J. Emneus and the COST MP1205 initiative.

Author Contributions: The optofluidic chip was designed by Marco Matteucci, Kirstine Berg-Sørensen, Rafael J. Taboryski, and Anders Kristensen. Marco Matteucci designed the shim and produced the optofluidic chip. Marco Triches and Mark R. Pollard performed the Raman experiments. Giovanni Nava and Kirstine Berg-Sørensen performed the optical stretching experiments. All authors contributed to the writing of the manuscript.

Conflicts of Interest: The authors declare no conflict of interest. The founding sponsors had no role in the design of the study; in the collection, analyses, or interpretation of data; in the writing of the manuscript, and in the decision to publish the results.

References

1. Psaltis, D.; Quake, S.R.; Yang, C. Developing optofluidic technology through the fusion of microfluidics and optics. *Nature* **2006**, *442*, 381–386. [CrossRef] [PubMed]
2. Fan, X.; White, I.M. Optofluidic microsystems for chemical and biological analysis. *Nat. Photonics* **2011**, *5*, 591–597. [CrossRef] [PubMed]
3. Schaap, A.; Bellouard, Y.; Rohrlack, T. Optofluidic lab-on-a-chip for rapid algae population screening. *Biomed. Opt. Express* **2011**, *2*, 658–664. [CrossRef] [PubMed]
4. Khoury, M.; Vannahme, C.; Sørensen, K.T.; Kristensen, A.; Berg-Sørensen, K. Monolithic integration of DUV-induced waveguides into plastic microfluidic chip for optical manipulation. *Microelectron. Eng.* **2014**, *121*, 5–9. [CrossRef]
5. Lai, C.W.; Hsiung, S.K.; Yeh, C.L.; Chiou, A.; Lee, G.B. A cell delivery and pre-positioning system utilizing microfluidic devices for dual-beam optical trap-and-stretch. *Sens. Actuators B Chem.* **2008**, *135*, 388–397. [CrossRef]
6. Ferraro, J.; Nakamoto, K.; Brown, C.W. *Introductory Raman Spectroscopy*, 2nd ed.; Academic Press: New York, NY, USA, 2003.
7. Nunes, P.S.; Ohlsson, P.D.; Ordeig, O.; Kutter, J.P. Cyclic olefin polymers: Emerging materials for lab-on-a-chip applications. *Microfluid. Nanofluid.* **2010**, *9*, 145–161. [CrossRef]
8. Khetani, A.; Riordon, J.; Tiwari, V.; Momenpour, A.; Godin, M.; Anis, H. Hollow core photonic crystal fiber as a reusable Raman biosensor. *Opt. Express* **2013**, *21*, 12340–12350. [CrossRef] [PubMed]
9. Østergaard, P.F.; Lopacinska-Jørgensen, J.; Pedersen, J.N.; Tommerup, N.; Kristensen, A.; Flyvbjerg, H.; Silahtaroglu, A.; Marie, R.; Taboryski, R. Optical mapping of single-molecule human DNA in disposable, mass-produced all-polymer devices. *J. Micromech. Microeng.* **2015**, *25*, 105002. [CrossRef]
10. Kou, Q.; Yesilyurt, I.; Studer, V.; Belotti, M.; Cambril, E.; Chen, Y. On-chip optical components and microfluidic systems. *Microelectron. Eng.* **2004**, *73–74*, 876–880. [CrossRef]
11. Kou, Q.; Yesilyurt, I.; Chen, Y. Collinear dual-color laser emission from a microfluidic dye laser. *Appl. Phys. Lett.* **2006**, *88*, 091101. [CrossRef]
12. Constable, A.; Kim, J.; Mervis, J.; Zarinetchi, F.; Prentiss, M. Demonstration of a fiber-optical light-force trap. *Opt. Lett.* **1993**, *18*, 1867–1869. [CrossRef] [PubMed]
13. Lincoln, B.; Schinkinger, S.; Travis, K.; Wottawah, F.; Ebert, S.; Sauer, F.; Guck, J. Reconfigurable microfluidic integration of a dual-beam laser trap with biomedical applications. *Biomed. Microdevices* **2007**, *9*, 703–710. [CrossRef] [PubMed]
14. Suresh, S.; Spatz, J.; Mills, J.P.; Micoulet, A.; Dao, M.; Lim, C.T.; Beil, M.; Seufferlein, T. Connections between single-cell biomechanics and human disease states: Gastrointestinal cancer and malaria. *Acta Biomater.* **2005**, *1*, 15–30. [CrossRef] [PubMed]
15. Ashkin, A.; Dziedzic, J.M.; Bjorkholm, J.E.; Chu, S. Observation of a single-beam gradient force optical trap for dielectric particles. *Opt. Lett.* **1986**, *11*, 288–290. [CrossRef] [PubMed]
16. Guck, J.; Ananthakrishnan, R.; Mahmood, H.; Moon, T.J.; Cunningham, C.C.; Käs, J. The Optical Stretcher: A Novel Laser Tool to Micromanipulate Cells. *Biophys. J.* **2001**, *81*, 767–784. [CrossRef]
17. PolyNano Summer School—Course Number 33692. Available online: http://www.kurser.dtu.dk/33692.aspx?menulanguage=en-gb (accessed on 22 October 2015).
18. Cubillas, A.M.; Unterkofler, S.; Euser, T.G.; Etzold, B.J.M.; Jones, A.C.; Sadler, P.J.; Wasserscheid, P.; Russell, P.S.J. Photonic crystal fibres for chemical sensing and photochemistry. *Chem. Soc. Rev.* **2013**, *42*, 8629–8648. [CrossRef] [PubMed]
19. Maples, R.E. *Petroleum Refinery Process Economics*, 2nd ed.; Pennwell Pub: Nashua, NH, USA, 2000.
20. Boyaci, I.H.; Genis, H.E.; Guven, B.; Tamer, U.; Alper, N. A novel method for quantification of ethanol and methanol in distilled alcoholic beverages using Raman spectroscopy. *J. Raman Spectrosc.* **2012**, *43*, 1171–1176. [CrossRef]

21. Numata, Y.; Iida, Y.; Tanaka, H. Quantitative analysis of alcohol—Water binary solutions using Raman spectroscopy. *J. Quant. Spectrosc. Radiat. Transf.* **2011**, *112*, 1043–1049. [CrossRef]
22. Tanzi, S.; Østergaard, P.F.; Matteucci, M.; Christiansen, T.L.; Cech, J.; Marie, R.; Taboryski, R.J. Fabrication of combined-scale nano-and microfluidic polymer systems using a multilevel dry etching, electroplating and molding process. *J. Micromech. Microeng.* **2012**, *22*, 115008. [CrossRef]
23. Matteucci, M.; Christiansen, T.L.; Tanzi, S.; Østergaard, P.F.; Larsen, S.T.; Taboryski, R. Fabrication and characterization of injection molded multi level nano and microfluidic systems. *Microelectron. Eng.* **2013**, *111*, 294–298. [CrossRef]
24. Jones, D.C.; Bennett, C.R.; Smith, M.A.; Scott, A.M. High-power beam transport through a hollow-core photonic bandgap fiber. *Opt. Lett.* **2014**, *39*, 3122–3125. [CrossRef] [PubMed]
25. Ashok, P.C.; Singh, G.P.; Rendall, H.A.; Krauss, T.F.; Dholakia, K. Waveguide confined Raman spectroscopy for microfluidic interrogation. *Lab Chip* **2011**, *11*, 1262–1270. [CrossRef] [PubMed]
26. Williams, G.O.S.; Chen, J.S.Y.; Euser, T.G.; Russell, P.S.J.; Jones, A.C. Photonic crystal fibre as an optofluidic reactor for the measurement of photochemical kinetics with sub-picomole sensitivity. *Lab Chip* **2012**, *12*, 3356–3361. [CrossRef] [PubMed]
27. Utko, P.; Persson, F.; Kristensen, A.; Larsen, N.B. Injection molded nanofluidic chips: Fabrication method and functional tests using single-molecule DNA experiments. *Lab Chip* **2011**, *11*, 303–308. [CrossRef] [PubMed]
28. Mucha, J.M.; Szytuła, A.; Kwiatkowska, C.J. Magnetic properties of nickel-vanadium alloys. *J. Magn. Magn. Mater.* **1984**, *42*, 53–58. [CrossRef]
29. Seryogin, G.; Golovato, S.; Smith, S.; Williams, K.; Limburn, N.; Winn, A.; Mundada, G.; Adema, G. NiV Stress Control Utilizing PVD with an Ar/N$_2$ Gas Mixture. In Proceedings of the 45th International Symposium on Microelectronics, San Diego, CA, USA, 9–13 September 2012; pp. 73–78.
30. Buck, J.A. *Fundamentals of Optical Fibers*, 2nd ed.; John Wiley & Sons: Hoboken, NJ, USA, 2004.
31. Antonopoulos, G.; Benabid, F.; Birks, T.A.; Bird, D.M.; Knight, J.C.; Russell, P.S.J. Experimental demonstration of the frequency shift of bandgaps in photonic crystal fibers due to refractive index scaling. *Opt. Express* **2006**, *14*, 3000–3006. [CrossRef] [PubMed]
32. Bland-Hawthorn, J.; Ellis, S.C.; Leon-Saval, S.G.; Haynes, R.; Roth, M.M.; Löhmannsröben, H.G.; Horton, A.J.; Cuby, J.G.; Birks, T.A.; Lawrence, J.S.; *et al.* A complex multi-notch astronomical filter to suppress the bright infrared sky. *Nat. Commun.* **2011**, *2*, 581. [CrossRef] [PubMed]
33. Guo, J.; Ma, X.; Menon, N.V.; Li, C.M.; Zhao, Y.; Kang, Y. Dual Fluorescence-Activated Study of Tumor Cell Apoptosis by an Optofluidic System. *IEEE J. Sel. Top. Quant. Electron.* **2015**, *21*, 7100107.

micromachines

MDPI

Review

A Comprehensive Review of Optical Stretcher for Cell Mechanical Characterization at Single-Cell Level

Tie Yang [1], Francesca Bragheri [2] and Paolo Minzioni [1,*]

[1] Department of Electrical, Computer, and Biomedical Engineering, Università di Pavia, Via Ferrata 5A, Pavia 27100, Italy; yangtie@gmail.com
[2] Institute of Photonics and Nanotechnology, CNR & Department of Physics, Politecnico di Milano, Piazza Leonardo da Vinci 32, Milano 20133, Italy; francesca.bragheri@ifn.cnr.it
*   Correspondence: paolo.minzioni@unipv.it; Tel.: +39-0382-985221; Fax: +39-0382-422583

Academic Editors: Shih-Kang Fan, Da-Jeng Yao and Yi-Chung Tung
Received: 10 March 2016; Accepted: 21 April 2016; Published: 13 May 2016

Abstract: This paper presents a comprehensive review of the development of the optical stretcher, a powerful optofluidic device for single cell mechanical study by using optical force induced cell stretching. The different techniques and the different materials for the fabrication of the optical stretcher are first summarized. A short description of the optical-stretching mechanism is then given, highlighting the optical force calculation and the cell optical deformability characterization. Subsequently, the implementations of the optical stretcher in various cell-mechanics studies are shown on different types of cells. Afterwards, two new advancements on optical stretcher applications are also introduced: the active cell sorting based on cell mechanical characterization and the temperature effect on cell stretching measurement from laser-induced heating. Two examples of new functionalities developed with the optical stretcher are also included. Finally, the current major limitation and the future development possibilities are discussed.

Keywords: single-cell analysis; microfluidics; optofluidics; optical stretcher; mechanical properties characterization

1. Introduction

During the last decades, the technologies that were initially developed and carefully optimized for microelectronic device fabrications widely expanded into other scientific research fields. One of the most relevant results of this "contamination" between microelectronic fabrication technologies and other research fileds led to the creation and development of the first microfluidic devices and lab-on-chip (LoC) systems. After their appearance in the 1990s, LoC systems have grown rapidly to become a very hot topic, due to the inherent advantages they offer with respect to "standard approaches", including miniaturization, parallelization, integration, automation, as well as low consumption, high efficiency, rapid analysis, cost-effectiveness. Compared with conventional methods, LoC devices offer a great potential and they enabled new biomedical applications, ranging from drug discovery and delivery, to disease diagnosis and point of care (POC) devices. Since the size-scale of LoC internal structures is of the same order of most cells' size, these devices have been extensively used for cellular biology studies, and in particular to analyze the biophysics and biomechanics of single cells [1–4].

Cell mechanical properties are mainly determined by the cellular cytoskeleton, which is a complex network of filaments, microtubules and linkers. It has been proved by many researches that cell mechanical properties are directly related to the cell status [5–7], like cell proliferation, differentiation and pathology transformation, particularly related to cancer. As an example, several studies demonstrated that cellular neoplastic and malignant transformations are closely connected with significant changes in the cytoskeleton, which are in turn related to changes in the mechanical properties of the cell [8–10].

Many different methods and experimental techniques have currently been proposed to assess cellular mechanical properties, either in a quantitative or qualitative way. For example, Vaziri, Pachenari *et al.* [11,12] applied a negative pressure in the micropipette to create an "aspiration region" on the cell and studied the local membrane deformation at the contact area; Mathur, Mackay, Rouven Brückner *et al.* [13–15] determined the local cellular Young's modulus or the cell plasma membrane tension by using an AFM cantilever tip on the cells' surface and measuring the relative indentation depth at constant force; Dao *et al.* [16] and Chen *et al.* [17] exploited optical tweezers or magnetic tweezers, with microbeads attached to the cell membrane, to apply a very large force onto the cell surface, and they derived the cellular viscoelastic moduli from the cell deformation. Preira, Luo, Martinez Vazquez *et al.* [18–20] developed a microfluidic chips with small constriction channels and applied them to the analysis of cell migratory capabilities, allowing to study both active and passive cell mechanical properties. However, some of these techniques can only access and hence probe a small portion of the cell, and most of them need a direct physical-contact between the studied cell and the device, which could modify cell's natural behavior and even damage it during the measurement. Furthermore, these techniques often require quite complicated experimental preparations and they offer a relatively limited throughput. Recently, Otto, Mietke *et al.* [21,22] developed a purely hydrodynamic cell-stretching technique that allows increasing significantly the measurement throughput; this method is ideally suited when large populations of cells are analyzed, but it doesn't allow cell recovery for further studies.

In contrast, the optical stretcher (OS in the following) proposed by Guck *et al.* [8] proved to be a very powerful tool for the study of cell mechanics: it is an optofluidic device combining the use of a microfluidic channel together with laser beams for optical stretching. The laser radiation applies a contact-less force on cell surface, causing a deformation that depends on cell mechanical properties. The use of a microfluidic integrated configuration allows achieving a high trapping (and analysis) efficiency of the cells flowing in the channel. Several studies already demonstrated that cell optical deformation measured from optical stretcher can be used as a mechanical marker to distinguish healthy, tumorigenic and metastatic cells, as well as to reveal the effects of drug treatments on the mechanical response of the cell [8,23–25].

In this paper we give a comprehensive review of the OS, including different fabrication techniques and materials, working mechanism and different applications. In addition, several new developments and findings from recent studies are also described.

2. Different Fabrication Techniques and Material

Thanks to the great improvement of micromachining technology, LoC and microfluidic device performance significantly advanced during the last decade. In this section we review the different materials and techniques that were reported in the literature for OS fabrication.

2.1. Basic Structure of an OS

The basic structure of an OS is schematically illustrated in Figure 1 and it is based on a dual-beam laser trap in a microfluidic circuit. The microfluidic network is typically composed by a single channel (even if multiple-input and multiple-output structures can be realized) allowing the cell suspension to flow from an external reservoir (e.g., a vial) to the laser trap and then to the output, which can be a sterile vial, or even a simple water drop. In order to achieve the best performance, the cross section of the channel should be rectangular, to avoid "lensing effects" from the channel-fluid interface, and the surface roughness should be extremely low, to allow a high imaging quality and to reduce the laser beam distortions at the interface. The laser trap should be designed and realized so that two identical counter-propagating beams cross the microchannel, generally in the "lower half" of the channel so as to easily intercept the cells flowing in the channel, e.g., 25 µm above the floor as reported in [26] , where cells with a typical dimension ranging from 5 to 20 μm are considered. The height of the flowing cells can be slightly modified by tuning the flow speed. It was experimentally found that a good height to position the optical trap is between 20 and 40 μm from the channel floor since it prevents the cells from depositing on the floor, while keeping the cells flowing slowly. Furthermore, the two laser beams should be preferably aligned perpendicularly to the flow direction, and they should be symmetrically positioned with respect to channel axis.

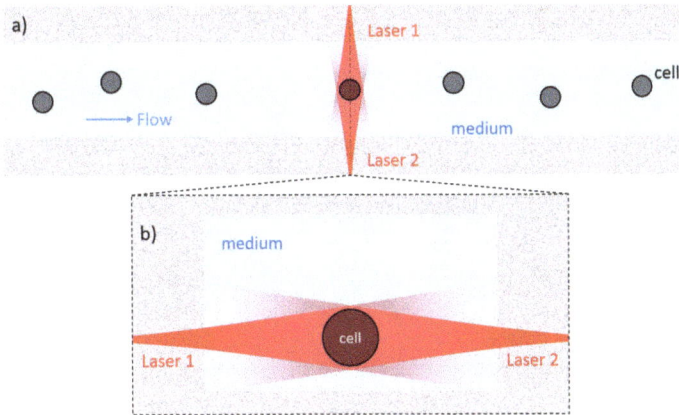

Figure 1. Schematic structure of an optical stretcher. (**a**) Top view of the central fluidic channel with flowing cells and a single cell trapped in the middle of the two opposing laser beams; (**b**) Cross section of the trapping area as indicated by the dash line in (**a**). The two opposite laser beams are positioned close to the channel floor.

Different LoC systems and fabrication techniques were proposed in the literature to realize an OS, including semiconductors, polymers and glasses, each of them having specific properties and hence allowing to integrate different features in the final device [27]. As an example silicon allows for surface stability and thermal conductivity, but it is opaque, hence undesirable for imaging purposes. Polymers, which can be biocompatible and transparent, offer the advantages of low cost and availability of simple technologies for microchannel fabrication, even if their hydrophobicity and softness may be a problem for some applications. Glass, on the other hand has the advantage of being chemically inert, stable in time, hydrophilic, nonporous and it easily supports electro-osmotic flow. Moreover, when fused silica is considered, it possesses a very high optical transparency range, down to the UV, and very low background fluorescence, surface coating can be easily performed and optical waveguides can be integrated in the substrate.In the following, we summarize the different methods for OS fabrication.

2.2. Conventional Discrete-Elements OS

Similarly to many optofluidic devices, the OS in its first implementation [8,28] was realized using discrete optical and fluidic components: two optical fibers were simply faced to a flow chamber where the cell suspension was flown. These first prototypes suffered from vibrations and mechanical drifts, which affect the system alignment and lead to barely repeatable experiments. To solve these problems, a solution reported in the literature is to align the optical and fluidic components on a substrate, by exploiting lithographically fabricated grooves, and then to seal the system with a suitable cover to increase the device robustness [26]. The fabrication procedure of such assembled optical stretcher (AOS) is illustrated in Figure 2a. A glass substrate is patterned with an SU-8 photoresist structure using standard photolithographic techniques. This leads to a single rectangular region of constant height (typically 35 μm) having perpendicular gaps to align and hold the optical and fluidic components.

In particular a square glass capillary is used to transport the cell suspension and two optical fibers, single mode at wavelengths >1 μm (Hi-1060, Corning, New York, NY, USA), are used to create the dual-beam optical trap. A thin slab of polydimethylsiloxane (PDMS) with a 1.5 mm hole is placed over the setup so that the trap region is centered within the hole. The hole is filled with index matching gel to reduce reflection of the laser beams. A glass coverslip is secured over the PDMS piece. Finally the PDMS layer is screwed into the microscope stage and the capillary is connected to the external tubing network. In Figure 2b, the finished system is placed over an inverted phase contrast microscope for cell imaging.

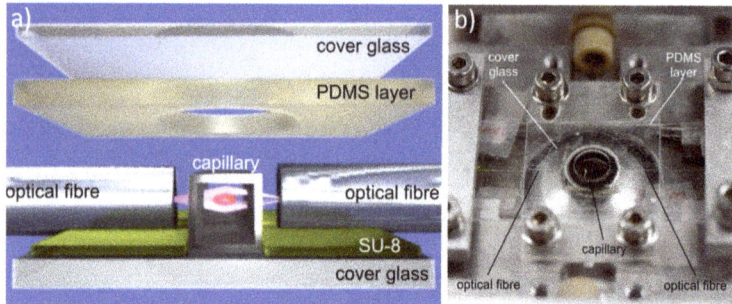

Figure 2. The structure of the discrete-element optical stretcher. (**a**) 3D rendering of the constituent components and their positions; (**b**) The finished system is mounted on a microscope plate. Figure reproduced from Reference [29] with permission from Optical Society of America.

2.3. Second-Generation Assembled OS

Two new methods to improve the tolerances of the AOS fabrication have been recently proposed. The first one is based on the use of two asymmetrically-etched glass substrates to accommodate both the flow channel and the fibers [25], as shown in Figure 3. The volumes etched from the top glass-layer include the majority of the optical fiber and the entire flow channel; on the other side, the bottom layer includes a shallow groove, for the fibers only. Through careful choice of the chip layout, good fiber alignment and cell trapping position can be achieved, together with a significant robustness to the misalignment of the two glass pieces. In fact, as shown in Figure 3, even a big misalignment of the top glass layer does not affect the flow channel and the fiber position. A small laser distortion is expected due to the etched curved surface, which can be minimized by fine tuning the geometry of the etching layout.

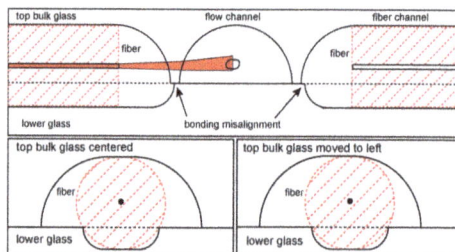

Figure 3. The chip geometry of the double glass layer assembled optical stretcher. The two glass layer are etched asymmetrically, the top one for large part of the fiber and the entire fluidic channel, the bottom one with shallow grooves for fiber alignment. The misalignment between these two layers show the robustness of this new etching layout. Figure reproduced from Reference [25] under CC-BY 3.0 license.

The second one is based on soft polymer material [30]. The proposed chip is fabricated by exploiting an innovative process, encompassing a double resist exposure and the use of Cyclin Olefin Copolymer (COC) TOPAS5013, which allows the fabrication of a multi-layer stamp shim. The two-level grooves for fiber positioning respect to the channel are obtained, as illustrated in Figure 4a. After the insertion of the optical fibers in the auto-aligning grooves they are glued and then the chip is sealed by thermal bonding of a TOPAS foil, see Figure 4b,c. An interesting feature of this approach is that different structure of the microfluidic channel can be easily obtained by changing the shim, and the easy chip fabrication method makes it ready for mass-production.

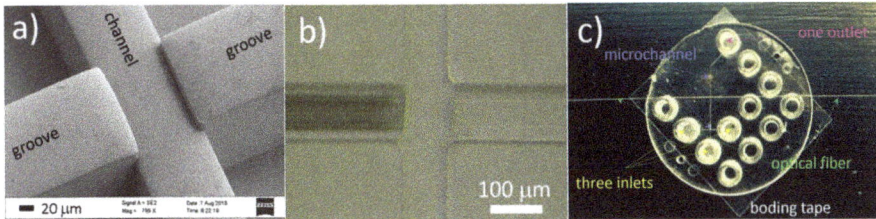

Figure 4. The polymeric assembled optical stretcher. (**a**) SEM image of the Ni shim at the area where the fiber grooves meet the channel. Higher position of the fiber grooves with respect to the central channel will lead to the lower position of the fibers after insertion; (**b**) Bright filed microscope image of the chip at the same area with a photonic crystal fiber (left) and single mode fiber (right) inserted; (**c**) The finished chip ready for use. The three inlets are for hydrodynamic focusing purpose. Figure reproduced from Reference [30] under CC-BY 4.0 license.

2.4. Femtosecond Laser Fabricated Monolithic Optical Stretcher

A completely different approach consists in the fabrication of a monolithic optical stretcher (MOS) by femtosecond laser micromachining (FLM) technology [29]. FLM holds many advantages over other fabrication techniques [31,32]: (i) it can be applied to different transparent materials; (ii) it enables rapid prototyping of devices; (iii) it is a 3D technique, allowing the fabrication of waveguides at different depths in the substrate; (iiii) it allows overcoming the complicated structure design and the assembly procedure. FLM has been extensively applied for a lot of optofluidic microchips in many studies. By the technique known as Femtosecond Laser Irradiation followed by Chemical Etching (FLICE) [33–36], the direct fabrication of microfluidic channels in fused silica can be obtained, which makes possible the integration of the microfluidic network and the optical waveguides by the laser radiation in the same substrate.

The first example of a MOS realized by FLM was obtained by integrating optical waveguides into a commercial microfluidic chip produced by Translume Inc. (Michigan, MI, USA). The chip is based on a "3-layers technology" where a fused silica glass slide with a thickness of 250 μm is machined with the FLICE technique to obtain a through slot as the fluidic channel [29,37], see Figure 5. The machined layer is then placed in the middle of the structure and sealed by thermal bonding with two polished fused silica glass slides on both top and bottom surfaces . In particular, while the bottom layer is a simple, unmodified, slide, the top layer has two through-holes aligned with the middle-layer channel terminations, so as to form the input and output accesses of the microchannel. The subsequent fabrication of pairs of opposing waveguides orthogonal to the channel was also realized by femtosecond laser writing, which allows adjusting the waveguides "depth" with respect to the channel during the laser writing process. With such a fabrication technology, the channel cross-section has a perfect rectangular shape with optical quality for the top and bottom channel walls and very low surface roughness of 200 nm rms for the lateral walls, thus allowing for a good imaging quality and a high efficiency optical-trapping by the dual beam configuration. Additionally the opposing waveguides are also aligned with a very high precision, thus allowing to obtain a robust, portable and highly flexible monolithic OS, see Figure 5c.

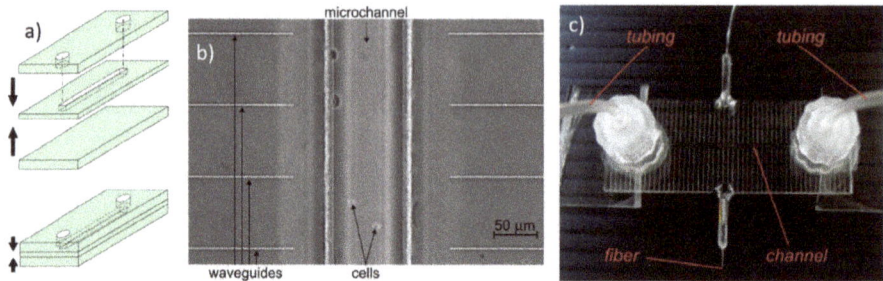

Figure 5. (**a**) Schematic representation of the three-layer technology for the monolithic optical stretcher fabrication. The central fused silica glass is machined by Femtosecond Laser Irradiation followed by Chemical Etching (FLICE) technique for the central channel and then sealed on both sides with two polished glass slides; (**b**) Microscope image of the straight microfluidic channel with pairs of waveguides beside it; (**c**) The finished chip is pigtailed with optical fibers and connected with external tubing through Luer connectors. Figure reproduced from Reference [29] with permission from Optical Society of America.

A different approach of monolithic OS, fully realized by FLICE in a single piece of silica glass, was demonstrated by Bellini *et al.* and Bragheri *et al.* [36,38]. Despite the significant advantages given by the possibility to realize in a single "writing procedure" both the microchannel and the optical waveguides, this method showed the significant drawback of producing microchannels with a high surface roughness, which can lead to a low image quality for the cell imaging. However, in a recent paper by Yang *et al.* [39] a new laser writing geometry allowing to decrease the surface roughness and to strongly improve the image quality was reported and further discussed in Section 5.1.

3. Working Principle of the OS

In this section we briefly review the basic physical principles underlying the mechanism of an OS, which is independently of its design, material and fabrication technique. In particular, we give a detailed description of the optical force distribution calculation, of the cell optical stretching procedure and of the method used to characterize cell deformation.

3.1. Optical Forces for Cell Stretching

As optical tweezers, also the OS exploits optical forces to both trap cells, but differently from tweezers, the scattering force in OS can be effectively used also to stretch the samples. Each photon from the laser beam carries a specific momentum, given by Equation (1), where h is the Planck constant; λ and v the wavelength and frequency of the light; c_0 the light speed in vacuum; and n the refractive index of the traveling material. This momentum is modified both in modulus and direction when the photon crosses through the interface between the external medium and the cell. To correctly evaluate the force distribution applied on the cell surface it is important to take into account that both reflection and transmission occur at the interface between the cell and the suspension medium; thus a different momentum-change is experienced by reflected and transmitted photons. The ratio between reflected and transmitted photons is given by Fresnel equations, which are generally applied under the hypothesis of unpolarized radiation, and hence depends on the angle formed between the photons propagation direction and the cell surface, as well as by the refractive indices of the medium and the cell (with the general assumption that $n_{cell} > n_{medium}$).

$$\left.\begin{array}{l} p = h/\lambda \\ c = c_0/n = \lambda\upsilon \end{array}\right\} \quad p = \frac{hn}{c_0\upsilon} \tag{1}$$

Since, up to now, no method allows a direct measurement of the optical force distribution on the cell surface, some mathematical and computational models have been proposed to calculate it. One of the main problems is to obtain a proper determination of the force applied by the impinging laser radiation on each point of the surface. Starting from the seminal works by Ashkin *et al.* [40–42], it is possible to evaluate the overall optical force applied on a dielectric sphere by an impinging optical beam thanks to its decomposition into a series of optical rays which are assumed to compose the optical beam (standard ray optics—SRO). This ray optics approach was then adapted by Guck *et al.* [8,28] for the optical force distribution on the cell surface inside the optical stretcher. And his model is valid because normally the studied biological cell samples have a much bigger size (~10 μm) than the wavelength of the applied laser light, which is called large particle regime. This approach is however not sufficient to describe the full laser-particle interaction when loosely-focused (or collimated) beams are considered, as in that case the interaction region generally lies within the beam Rayleigh-range, and thus a different beam-decomposition technique (paraxial ray optics—PRO, [43]) has to be applied. Afterwards, some other studies improved this method by including the effects of multiple internal reflections of the laser light inside the cell [44–46].

In the following we briefly review the description of the PRO approach, so as to give the reader a better understanding of the beam decomposition and optical force calculation technique. The optical force calculation is basically performed as a two steps process: first an adequate decomposition of the optical beam as a series of optical rays in the region of interest (*i.e.*, in the area occupied by the cell/particle) is calculated, and then the interaction of each optical ray with the cell/particle is calculated, thus allowing to evaluate also the overall beam effect on the sample.

The first step, as shown an sketch in Figure 6a, is to decompose a non-focused Gaussian laser beam (as that generally emitted by optical fibers) into a distribution of individual rays, each characterized by a tproper direction and position in space, and carrying a certain amount of optical power. Considering the far field distribution through an angular spectrum decomposition technique, the power carried by a ray that intercepts the coordinate z at a distance ρ_0 from the axis of the beam is calculated as the integral of the intensity as a function of the radial coordinate ρ, in the portion of the annulus delimited by $\Delta\rho$, see Figure 6a. The spatial phase gradient is then used to determine the propagation direction of each ray, which is perpendicular to the wavefronts. In particular, by adopting the paraxial approximation, and considering a Gaussian beam with waist w_0 at $z = 0$, the optical field can be analytically described by two simple equations giving the electric-field amplitude (A), and radius of curvature (R) of the wavefronts respectively

$$A\left(\rho, z\right) = A_0 \frac{w_0}{w_z} \exp\left(-\frac{\rho^2}{w_z^2}\right) \tag{2}$$

$$R\left(z\right) = z + \frac{z_R^2}{z} \tag{3}$$

where z_R (the so called Rayleigh range) and w_z (beam width as a function of propagation distance) are:

$$z_R = \frac{\pi w_0^2 n}{\lambda} \tag{4}$$

$$w_z = w_0 \sqrt{1 + \left(\frac{z}{z_R}\right)^2} \tag{5}$$

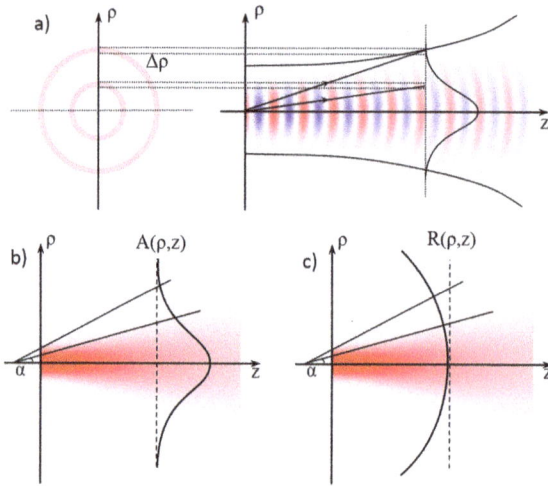

Figure 6. Scheme of the optical field determination from a Gaussian laser beam with the paraxial ray optics approach. (**a**) The power carried by each ray, is calculated as the integral of the beam intensity, as a function of the radial coordinate within the area of the annulus associated to the ray; (**b**) The amplitude $A(\rho, z)$ and the (**c**) curvature radius $R(\rho, z)$ along the axis z are calculated, which will be exploited respectively for the evaluation of the power and the propagation direction of each ray.

The amplitude of each ray is used to assess the optical power carried by each ray P, whereas the curvature radius is used to determine its propagation direction, thus obtaining a precise description of the beam properties in the area occupied by the cell/particle.

Then, as a second step, the interaction of each single ray with the cell/particle is considered. The reference system used to evaluate the interaction between each ray and the particle is shown in Figure 7a. We assume for the calculations that the particle has refractive index (n_p) larger than that of the surrounding medium (n_m). According to the analysis already reported in the literature by Ashkin *et al.* and Brevik [42,47], it it possible to evaluate the force transferred by each single optical ray because of its interaction with the particle surface, due to the difference of the refractive indices, as it undergo multiple reflections and refraction on the boundary of the sphere, see Figure 7b.

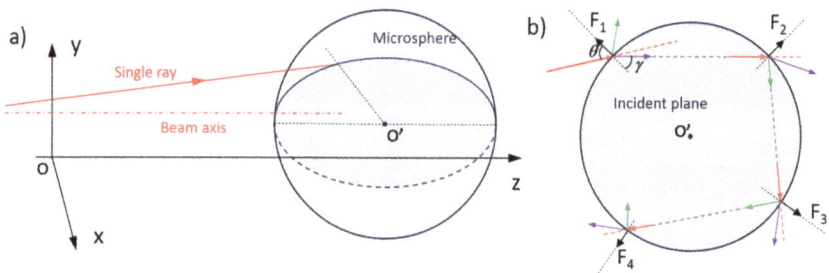

Figure 7. Coordinate system for a single ray interacting with a particle. (**a**) A single rays from a Gaussian laser beams hits on the surface of a particle. The locations of the laser beam and particle can be random. The incident plane is defined by the ray and the normal direction of the particle at the hitting point and is indicated by the gray area; (**b**) the single ray undergoes multiple reflections and refractions on the boundary of the particle.

Taking the first ray-particle interaction as an example, the incident angle is defined by the geometries of both the Gaussian laser beam and the particle and is referred as θ. According to Snell's law and Fresnel equations, the refractive angle γ and the transmission coefficients (T) and reflection coefficients (R) can be calculated exactly. Then, the momentum carried by the photons in the incident, reflection and refraction beams can be calculated through Equation (1). We denote the momentum of the incident, transmitted and reflected rays by $\vec{p}_i$, $\vec{p}_t$ and $\vec{p}_r$, and their directional unit vectors by $\vec{a}_i$, $\vec{a}_t$ and $\vec{a}_r$ respectively. According to the momentum conservation law, the stress $\vec{\sigma}$ applied on the local surface of the particle is expressed as the following equation [44]:

$$
\begin{aligned}
\vec{\sigma} &= \frac{\Delta \vec{p}}{A \Delta t} \\
&= \frac{\vec{p}_i - (\vec{p}_t + \vec{p}_r)}{A \Delta t} \\
&= \frac{n_m}{c_0} \frac{P}{A} [\vec{a}_i - (\frac{n_s}{n_m} T \vec{a}_t + R \vec{a}_r)] \\
&= \frac{n_m}{c_0} \frac{P}{A} \vec{Q}
\end{aligned}
\tag{6}
$$

where A is the area of the particle irradiated by the single ray, P is its optical power, c_0 is still the light speed in vacuum and $\vec{Q}$ is defined as a dimensionless momentum transfer vector. Furthermore, it is proved that the direction of this optical stress is always perpendicular to its acting surface and pointed away from the optically denser part [8,44], as shown in Figure 7b. In the same way, the optical stress from the following interaction of this single ray can be calculated and this can be easily adapted to other single rays by only changing the first incident angle and the associated power. The sum of the optical force from each ray results in a optical force distribution on the surface of the particle.

As an example, Figure 8 shows the calculated optical force distribution on a particle under different conditions. The applied Gaussian laser beam has a wavelength of 1.07 µm, a beam waist of 3.1 µm and carries an optical power of 10 mW. The particle, which is considered to have a diameter of 5 µm and to lay 30 µm away from the beam waist position, exactly along the laser beam axis, has refractive index of 1.37, similarly to real cells, and the surrounding medium has refractive index of 1.33 (corresponding to that of water). The optical stress distribution produced on the particle surface by a single laser beam shining from left to right and from right to left is shown in Figure 8a,b respectively. The resulted optical stress profiles are rotationally symmetric with respect to the beam axis and there is net force from a single laser beam pushing the particle away from the laser source. It is interesting to notice that the "spikes" which are observed in the stress profile are produced by the second order internal laser-particle interaction [44], and they tend to become more relevant as the particle diameter becomes smaller with respect to the beam one. When the two counter-propagating beams are simultaneously impinging on the particle, see Figure 8c, the net applied force is zero, and the particle is stably trapped in the center. However, the local optical force acting on the surface is not zero and it significantly increases with respect to the single laser case [8]. This situation is that commonly exploited to create an OS, by simply increasing the laser beams power. In Figure 8d–f we show the optical stress distributions of different particle size (5 µm, 10 µm, 15 µm) under the same situation of Figure 8c. Figure 8d is the same as Figure 8c and it is shown only to help the comparison. By comparing the three panels of the second line, it can be immediately seen that as the particle size is increased the optical force becomes more concentrated along the laser beam axis, and the stress distribution-shape slightly changes. Figure 8g–i show the optical stress distribution on a 5 µm particle when the distance of the particle from the beams waist is set to 30, 40 and 50 µm, respectively. By increasing the distance, the optical stress distribution slightly changes, and the overall stress is decreased, as can be intuitively understood noticing that the overall intercepted power is reduced as the beam broadens during propagation because of diffraction.

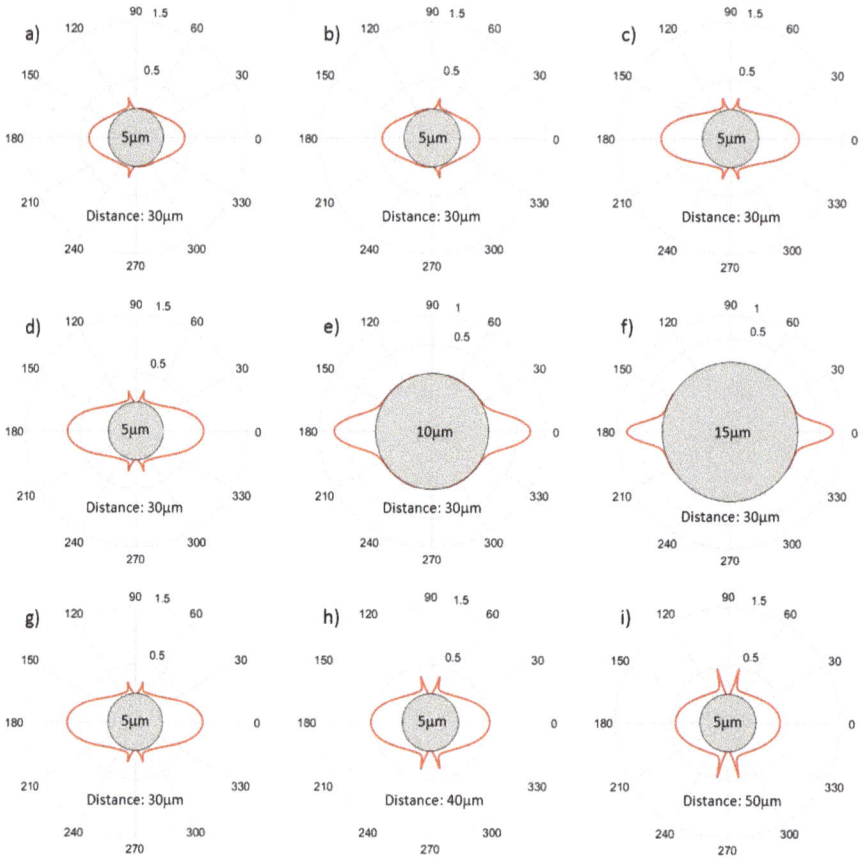

Figure 8. Polar plots of the calculated optical stress distribution on the surface of the particle. The Gaussian laser beam has beam waist of 3.1 μm and wavelength of 1.07 μm and carries optical power of 10 mW. The refractive index of the particle is 1.37 and medium 1.33. The distance between the particle center and the beam waist (either one laser or two lasers) is indicated in each panel together with the diameter of the particle. (**a–c**) show the optical stress from left side laser radiation, right side laser radiation and both side laser radiation respectively; From (**d**) to (**i**), the two-side laser irradiation is considered; (**d–f**) show the optical stress distribution change with particle size increase; (**g–i**) show the optical stress distribution change with distance increase.

3.2. Cell Stretching Procedure

Similar procedures for cell stretching measurements are reported in the literature [8,23,24,29] and the work-flow can be graphically depicted in Figure 9. First, the cell sample is injected by external pump system into the microfluidic circuit of the OS and low laser power is turned on for the two opposing laser beams. The cell flow is then adjusted as fast as possible, provided that the laminar flow is maintained and both a good imaging of flowing cells (which depends on the camera frame rate) and an efficient trapping (which depends on the cell velocity with respect to the laser power) are assured. Once a single cell reaches the laser irradiated zone and is trapped by the optical force, the flux is stopped, so that no additional cells reach the "trapping area". After this, the optical power output by the two waveguides is increased to the preset "stretching power" value with a step-like power profile and the cell progressive deformation is recorded by a camera. After a certain time interval (typically about 5 s), the laser power is switched back to the low "trapping" value, and the stretched cell is observed while it partially recovers the original shape. During the stretching and recovery processes, the images are recorded and saved at a high frame rate, to allow for subsequent image analysis. Finally, the studied cell is released, by switching off the laser beams and restarting the flux, and the whole procedure is then repeated for other single cells (additional details can be found in [8,23,29]).

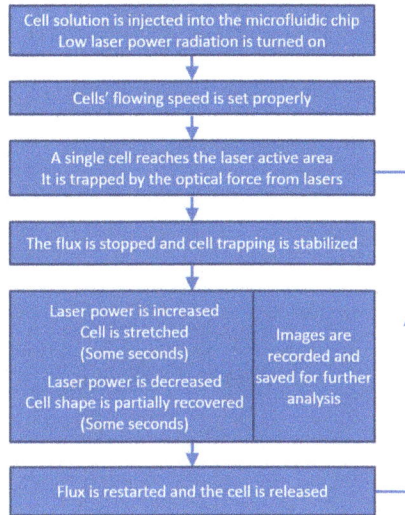

Figure 9. Flow chart of continuous cell stretching procedure.

3.3. Characterization of Cell Optical Deformation

In order to characterize the cell optical deformation from stretching measurements, the previously stored images are analyzed. There are different methods reported in the literature [8,26,48]. As an example, here we show the method, exploited by Guck, Lincoln, *et al.* [8,26], based on the image polar transformation and edge detection algorithms. The procedure is schematically represented by the different steps shown in Figure 10. First of all, each image is transformed into a polar coordinates system, by exploiting an automatic technique for cell-center identification. The radius limit of the polar coordinate is determined by the original rectangular image border, see the blue circle in Figure 10a, which is also present in Figure 10b as a straight line after polar transformation. Accordingly, the bottom part of Figure 10b is the outside of the cell.

The intensity of the gray scale image for each angular direction (*i.e.*, each vertical line in panel (b)) is extracted (see Figure 10c showing the intensity profile corresponding to the green line in Figure 10b) and smoothed by a low-pass Fourier filter to remove the small spikes. Then, the derivative of the intensity values along the radius is calculated and the minimum point (corresponding to the fastest transition from bright to dark region around the cell border) is recorded, as indicated by the red circle in Figure 10d. For each angle, the same process is applied and the cell border at each polar angle is obtained, see Figure 10e. Then this cell border curve in polar coordinates is transformed back and the cell contour is plotted upon the original phase contrast image as shown in Figure 10f. For the edge detection, some other signal processing methods can be additionally implemented to optimize the detection result, like squaring the intensity profile to have a stronger contrast or defining a threshold to simplify the global minimum search as demonstrated by Guck *et al.* [8]. With this method, a very high resolution of 100 nm can be obtained, which can satisfy the purpose of cell border recognition.

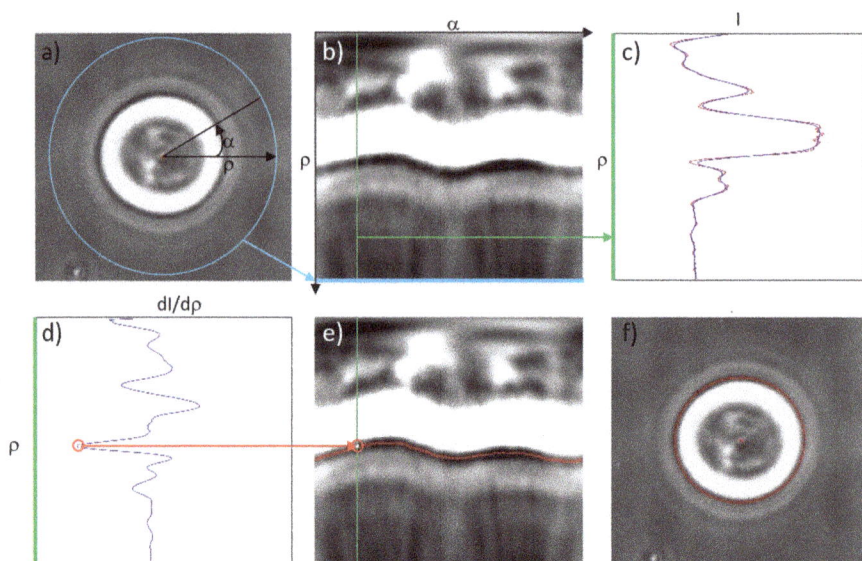

Figure 10. Illustration of the image analysis steps. (**a**) Phase contrast microscope image of a single cell. The center point (red color) of the cell is manually selected and the circular border (light blue color) for the polar transformation is determined by the original rectangular image border; (**b**) The polar-transformed image; (**c**) shows a plot of the gray-scale intensity along the green vertical line appearing in (**b**). The raw data (red line) is smoothed by Fourier filtering (blue line); (**d**) shows the first derivative of the blue-line (intensity) and the red circle shows the point identified as minimum, identified as belonging to cell border; (**e**) by repeating the same procedure for all the polar angles, the cell border is reconstructed on the polar image and (**f**) then transformed back to original image.

The image analysis procedure is applied to all the images from the cell optical stretching measurement and the cell contour is found in every image frame. Afterwards, the cell optical deformation in terms of elongation along the laser beam axis and contraction in the perpendicular direction can be derived from these cell contours. As an example, Figure 11 shows the deformation profile of a single cell (MCF7) under 5 s optical stretching and further 5 s recovery. For the stretching, a typical creep compliance curve of cell size variation can be observed. The cell deformation is reported as the absolute cell size value. The cell is increasingly elongated ($X_{elongation}$) along the laser beam axis and contracted ($Y_{contraction}$) in the perpendicular direction under stretching duration. After stretching, the cell can partially recover its shape. Figure 11c,d show the microscope images of the original cell trapped at the very beginning of the stretching and maximally stretched after 5 s of high-power irradiation.

To explain this time dependent and compliance behavior of the cell deformation in response to a constant step stress from the optical stretching measurement, Wottawah *et al.* [49,50] applied constitutive equations and, by fitting the experimental result with the theoretical prediction, the viscoelastic parameters of the cell, like Young's modulus or shear modulus, can be derived. Some other studies proposed more complicated mechanical 3D models to mimic the real complex structure of a single cell and study the different contribution of cellular mechanical structure. Ananthakrishnan *et al.* [51,52] developed two structural models: a thick shell model for the actin cortex and a three-layered model for the whole cell, and they found that the outer actin cortex mainly determines the structural response of the cell during cell stretching. Another interesting result was obtained by Gladilin *et al.* [53], which created a three-component model (including the nucleus, the cortical actin filament and the perinuclear vimentin intermediate filament) allowing them to isolate the contribution of each component, thanks to the use of specific drug treatments.

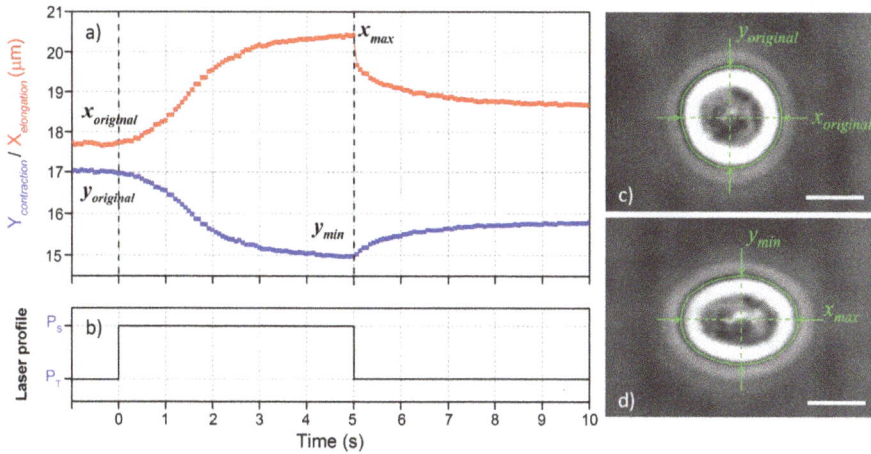

Figure 11. Single cell optical stretching. (**a**) Cell dimension variation during optical stretching together with the laser power profile (**b**): P_T is trapping laser power of 25 mW per side and P_S is stretching power of 1.5 W per side. X-axis is along the laser beam and Y-axis along the cell flowing direction; (**c,d**) show the phase contrast microscope images of the same cell trapped and stretched respectively. Green contours are cell borders identified by the recognition algorithm. Scale bars in both (**b,c**) are 10 μm. The cell sample is human breast cancer cell MCF7.

Besides, the cell maximum relative deformation (*i.e.*, strain ϵ) is also often used in literature as a simple reference parameter for cell mechanical property comparison. It can be either the relative elongation (Equation (7)) or the relative eccentricity variation (Equation (8)). In both equations, there is a correction term "*corr*" obtained by numerical simulations, which accounts for the variations of the optically induced stress profile from different cell size and the refractive index of the different cells as described in reference [8,23].

$$\epsilon(\%) = (x_{max}/x_{original} - 1) \cdot 100 \cdot corr \tag{7}$$

$$\epsilon(\%) = (\frac{x_{max}}{y_{min}} / \frac{x_{original}}{y_{original}} - 1) \cdot 100 \cdot corr \tag{8}$$

4. OS as a Tool to Analyze Cell Lines, Drug Treatments and Cellular Organelles

Since its invention, the OS clearly demonstrated its ability to test cell mechanical properties, and during the last decade, it has also been extensively and successfully applied to study different cell samples, the effect of various drug treatments on cell mechanics and internal cellular organelles.

4.1. Optical Stretching of Red Blood Cells and Lipid Vesicles

The first application of the OS to the study of single cell mechanics is related to the study of red blood cells (RBCs) [8]. The simple structure of RBCs (they do not have any organelles or internal structure), and the possibility to make them almost perfectly spherical by swelling in a hypo-tonic suspensions makes them to be the ideal candidate for preliminary analysis, as they can be effectively modeled as homogeneous spheres, with an isotropic index of refraction, thus simplifying the analysis of cell mechanical response. In Figure 12a, it is shown the microscope images of single RBC stretched at gradually increased laser power levels [45]. The cell is elongated along the laser beam axis and contracted in the perpendicular direction. Given the optical field of the stretching laser (magnitude and direction), the RBC size and its position in the optical field, the refractive indices of the cell and the buffer medium, the optical force applied on the cell surface can be precisely evaluated, as discussed in Section 3.1. The RBCs' relative deformation is represented in Figure 12b and it can be seen that the deformation of RBCs maintain a linear response with respect to the optical power until 150 mW. Within this linear regime, linear membrane theory can be used to describe the deformation of RBCs [8,45] (see the curve in the figure) and a value of Young's modulus about $Eh = (20 \pm 2)\mu Nm^{-1}$ was derived by Bareil *et al.*

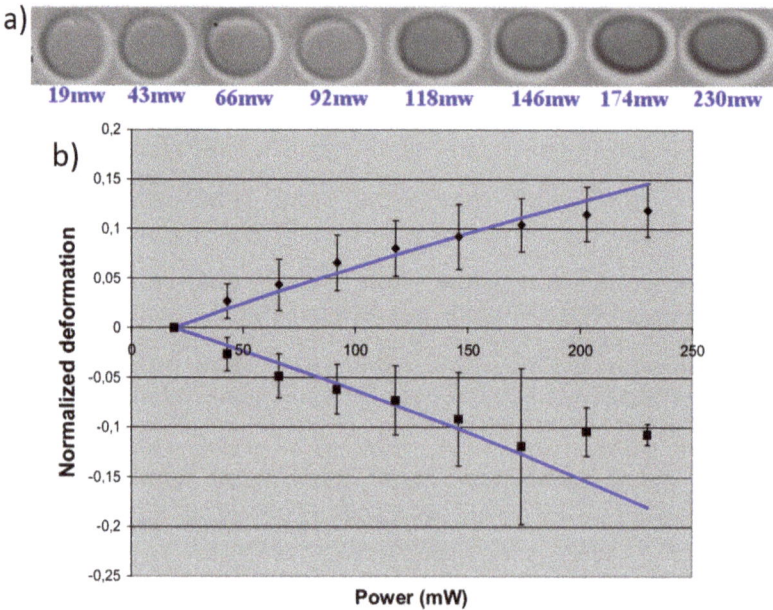

Figure 12. Optical stretching of single red blood cell (RBC). (**a**) Microscope images of RBCs stretched at increasing optical powers; (**b**) Optical deformation of RBCs in terms of elongation along laser axis and contraction in the perpendicular direction at different stretching power. The laser power is the total power for both laser beams. Experimental data is fitted with theoretical prediction from the linear elastic membrane theory. Figure reproduced from Reference [45] with permission from Optical Society of America.

Afterwards, many other related studies on optical stretching of red blood cells have been performed: Bareil, Ekpenyong *et al.* [44–46] calculated the exact stress distribution on the cell surface for a more accurate derivation of the cell stiffness; Ye, Mauritz, Sawetzki *et al.* [54–56] studied and compared the mechanical property of healthy *vs.* malaria-infected RBCs; Sraj *et al.* [57,58] applied laser diode-bar for RBCs optical stretching and proved a single-beam, high-throughput method for cell deformation cytometry; Liu , Tan *et al.* [59,60] developed a mechanical model and conducted a systematic simulation of RBCs dynamic deformation in optical stretching experiment.

Other studies showed the analysis of phospholipids vesicles (synthetic structures with a spheroidal lipid-bilaryer shape) by exploiting the OS measurement technique [50,61,62]. Similar to RBCs, they can provide a simple mechanical model for studying membrane mechanics, which is a very important aspect of the cellular function. In Figure 13, an example of a vesicle trapped at low laser power and then stretched at high power is reported. The time resolved deformation of vesicles under different laser power is plotted in Figure 13c. It is clear that vesicles have a fast response to the applied optical stimulus as they can reach their final deformation immediately after the stretching started and recover the deformation and return to the initial shape very rapidly after the stretching stopped. Interestingly, by stretching vesicle, Solmaz *et al.* [61] were able to extract the bending modulus of the lipid bilayer, while Delabre *et al.* [62] proposed a vescicle model based on a quasi-spherical approximation allowing also to take into account the laser heating effect on the vescicle deformation.

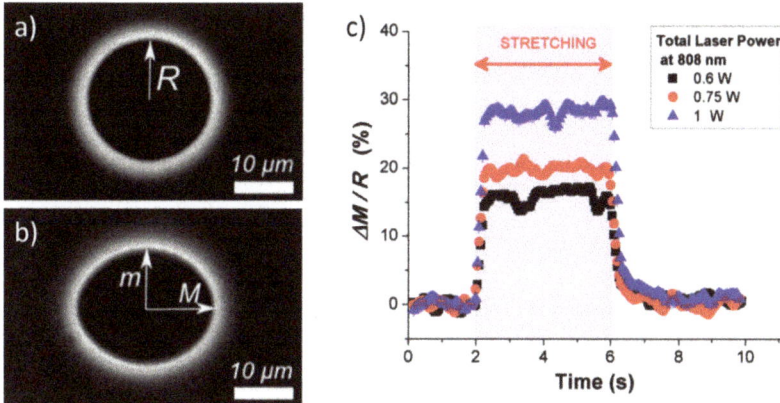

Figure 13. Optical stretching of vesicles. Microscope image of a vesicle trapped at low power (**a**) and deformed at high power (**b**); (**c**) The major axis strain of vesicles under 4 s stretching at various total powers. Figure reproduced from Reference [62] with permission from Royal Society of Chemistry.

4.2. Optical Stretching of Eukaryotic Cells and Drug Treatment Effect

Differently from erythrocytes, eukaryotic cells have a much more sophisticated structure including various internal organelles, nucleus and several immersed polymer networks. In particular, these polymer networks, which consist mainly of three different filamentous proteins: actin filaments, microtubules, and intermediate filaments, form the cellular cytoskeleton and provide the basic mechanical support for the whole cell in terms of mechanical strength and morphology [23,50]. Guck *et al.* [8] first exploited the optical stretcher to analyze BALB 3T3 fibroblast cells and showed that even if a very high laser power was applied, the cell showed a very small deformation. Because of its internal structure, BALB 3T3 cells behaved like a relatively hard sphere. Similar behaviours are obtained for different populations of eukaryotic cells. Figure 14 shows an example of a single MCF7 cell trapped at low power of 25 mW per side and stretched at much higher power of 650 mW per side; even in this case only a small deformation can be seen.

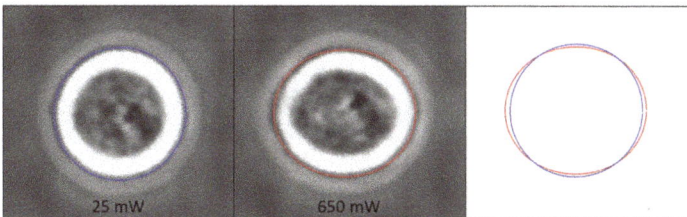

Figure 14. Optical stretching of a single MCF7 cell. The laser power is for each side and the cell contour is recognized by the edge detection algorithm in Figure 10.

After the successful application of the optical stretcher to eukaryotic cells, Guck *et al.* [23,26] exploited it to evaluate variations in the mechanical properties of cell lines at different evolution stages, e.g., the healthy cells with respect to the tumorigenic ones and even the metastatic ones. The effects of drug treatments on cell mechanical property have been also investigated. In particular, a well characterized cell line of human breast epithelial cells and their cancerous counterparts were considered by Guck and coworkers: MCF10, MCF7 and MDA-MB-231, which are respectively normal, cancerous and highly metastatic cells. In Figure 15, the optical deformability of the three cell lines is reported. The results showed that the curves of the cells' optical deformability can be fitted with a normal distribution and the obtained optical deformability values are: 10.5% ± 0.8% for MCF10, 21.4% ± 1.1% for MCF7 and 33.7% ± 1.4% for MDA-MB-231. With few cells measured, their optical deformability can be surprisingly distinguishable in a statistic way. The differences in the optical deformability of the three cell lines can be directly related with their different metastatic potentials.

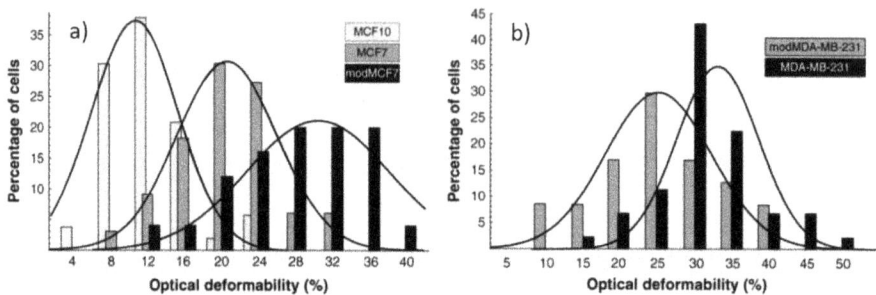

Figure 15. Optical deformability of normal, cancerous, and metastatic breast epithelial cells. (a) The three populations of the MCF cell and (b) the two populations of the MDA-MB-231 cell are clearly distinguishable. Curves represent the fitting of normal distribution. Figure reproduced with permission from Reference [23] with permission from Prof. Guck.

Besides, they applied trug treatments to the two cell lines of MCF7 and MDA-MB-231 and their results are also included in Figure 15. For MCF7 cell sample, the phorbol ester TPA was applied, which can dramatically increase MCF7 cell invasiveness and its metastatic potential; it can be observed in Figure 15a that the optical deformability is increased with this drug treatment (modMCF7) highlighting higher optical deformability corresponds to the higher metastatic potential. While, MDA-MB-231 cells treated with alltrans retinoic acid, became less aggressive and its optical deformability was decreased, see (modMDA-MB-231) in Figure 15b, which showed the reverse trend of metastatic competence with cell optical deformability.

Other cell lines have been also evaluated, Lautenschläger *et al.* [63] studied acute promyelocytic leukemia (APL) cells with optical stretcher and revealed a significant softening during differentiation. Furthermore, they exposed the cell to paclitaxel and found out that this treatment does not alter cells' compliance from optical stretching but reduces cell relaxation after the optical stress is removed. Schulze *et al.* [64] exploited the optical stretcher on human skin fibroblast cells and showed that an increase in age was clearly accompanied by a cell stiffening. Ekpenyong *et al.* [9] evaluated the differences in cell mechanical properties during differentiation of human myeloid precursor cells into three different lineages and observed that a reduction in steady-state viscosity is a physiological adaptation for enhanced migration through tissues.

5. Active Mechanical Sorting Based on Cellular Optical Deformability

As previously discussed, cellular optical deformability, measured by OS, provides a simple marker for cells analysis, allowing to distinguish healthy, tumorigenic and metastatic cells, as well as to study the cell mechanical response from different drug treatments [8,23,63]. Starting from this evidence, the possibility to use cells' optical deformability as a criterion for single-cell sorting was recently demonstrated, by two separate studies [24,25], thus opening the way to biological analysis requiring the selection and recovery of those cells that exhibit specific mechanical properties. The concept behind this result is quite simple: thanks to a real-time analysis of cell stretching (which can be realized by an automated computer process), the trapped and stretched cells are immediately recognized as "interesting or not", and the presence of two separate outputs in the microfluidic structure allows sorting the trapped cell according to the stretching measurement result, without requiring any additional marker, like fluorescent stain [65], and without needing a large cell population [66–69].

5.1. Chip Layout for Cell Stretching and Sorting

Cell sorting naturally requires more than one output, the microfluidic design has to be obviously different with respect to that of a standard OS, and also the optical section requires some modification. As an example, the optofluidic microchip proposed by Yang *et al.* [24] for active cell sorting on the basis of optical deformability is shown Figure 16 . This chip is realized in a very small piece of silica glass by the FLICE technique, as described in Section 2.4 and its size is 2 mm (thickness) × 1.5 mm (width) × 4 mm (length).

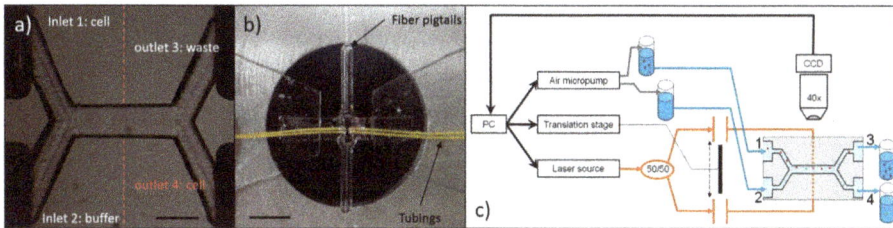

Figure 16. Active cell sorting chip design. (**a**) Microscope image of the internal structure of the cell sorting microchip. Scale bar: 100 µm; (**b**) The finished chip with fibers pigtailed and tubing connected is very compact. Scale bar: 1 cm; (**c**) Schematic of the experimental setup. Figure reproduced from Reference [24] with permission from Royal Society of Chemistry.

The microfluidic network design is almost the same of an optical sorter [65] and consists of an X-shaped microfluidic circuit, including two inlets, two outlets and a common central channel, see Figure 16a. The experiment setup is schematically shown in Figure 16c. After injecting cell suspension (from inlet 1) and buffer medium (from inlet 2), a laminar flow regime is built up in the central channel, allowing to keep the flowing cells in their own flow stream up to the chip outlet (port 3). Two fiber-to-fiber U benches, inserted along the optical section, provide the possibility of performing both cell stretching measurement and subsequent sorting procedure by optical forces with the same optical waveguides, by simply reducing the intensity of one beam (e.g., putting an attenuator in the corresponding U bench) and thus unbalancing the optical power levels between the two waveguides, so that the stronger beam will gently push the cell towards the desired stream flow, thus achieving cell sorting. When a cell is ready to be tested, a customized LabVIEW program is applied for the real-time analysis of optical stretching measurement; the obtained cell-deformation value is then automatically compared with a user-defined threshold, and the cell will be sorted accordingly, either to the "waste outlet" (port 3) or to the "selected cells" one (port 4). This procedure is continuously performed until the desired number of cells is sorted.

Meanwhile, another study from Faigle *et al.* [25] presented similar design but with different fabrication technique. Both the internal microfluidic channel and the optical fiber slots are created in separate glass substrate by chemical etching. The chip assembling, especially for fiber insertion and alignment, is improved by optimizing the etching and bonding geometry. The direct usage of optical fibers provides a high efficient optical power delivery from the laser source to the active stretching area and also the thin glass slide as the chip bottom layer guarantees a good imaging quality. Besides, the possibility of having more complex channel geometries can be obtained with this method.

The microchips generally realized by the FLICE technique have the big advantage of being monolithic and compact, however, the internal surface of the microfluidic channel may be quite rough, thus causing imaging distortion and hindering imaging quality, which is fundamental for an appropriate cell contour recognition and for a correct deformation evaluation. This issue was recently solved by exploiting a new laser irradiation geometry [24], as shown in Figure 17. The main idea is to write the microchannel structure "along" the beam propagation direction, and not perpendicularly to it, as in previous fabrications. This, apparently small, change in the fabrication procedure allows defining all the microchannel surfaces in a significantly smoother way, thus strongly decreasing the internal surface roughness and greatly improving the imaging quality, as evident in Figure 17. The physical reason underlying this surface quality improvement was suggested to be connected to the highly ellipsoidal shape of the writing voxel: exploiting the newly proposed approach, all the microchannel surfaces are produced by the "longer side" of the voxel, while the "shorter and sharper" part of it has almost no impact in the structure determination, because of the voxel-translation direction.

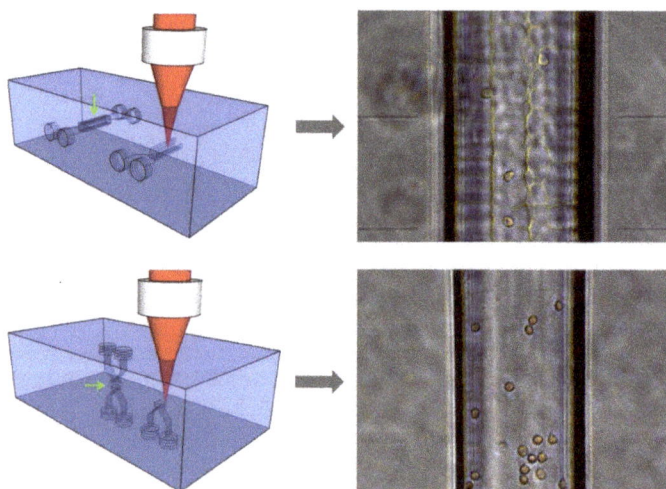

Figure 17. Laser writing geometry optimization for internal channel surface roughness control. Figure reproduced from Reference [24] with permission from Royal Society of Chemistry.

The physical reason underlying this surface quality improvement was suggested to be connected to the highly ellipsoidal shape of the writing voxel. By using the "longer side" of the voxel and keeping the same separation between the different "writing tracks", a larger overlap and a better uniformity of laser irradiation are therefore obtained leading to a smaller roughness after chemical etching.

5.2. Cell Sorting Efficiency Discussion

With the new proposed microchip, Yang *et al.* performed the cell sorting experiments with metastatic (A375P) and highly metastatic (A375MC2) human melanoma cells, two cellular lines with a very similar cell size distribution (17 ± 2 μm in diameter, see Figure 18) and a slightly different optical deformability (8.4% ± 1.1% for A375P and 10.1% ± 1.8% for A375MC2 using two optical beams of 1.2 W each), thus reflecting their different mechanical properties and offering an intrinsic cell marker to separate them. As discussed in Section 4.2, the higher optical deformability of A375MC2 is directly related with its higher metastatic potential. In addition, it should be noted that the distributions of the cell size and optical deformability from these two cell samples follows very well a normal distribution, see the fitting Gaussian curves in Figure 18.

The cell sorting experiment was carried using a 1:1 cell mixture of A375P and A375MC2 under the same concentration (so that the same number of A375P and A375MC2 cells is present in the final suspension), thus allowing to estimate that the overall deformation distribution could be considered as the sum of the two Gaussian curves reported in Figure 18 . After each cell stretching-measurement, the produced deformation was compared with a preset threshold value, and if the measured deformation was higher than the threshold, the cell was sorted into the collection branch of outlet 4; otherwise, the cell was addressed to the waste part of outlet 3, see Figure 16. In this way, an enriched sub-population of highly metastatic cell A375MC2 can be obtained, as shown an example in Figure 18c, by selecting cell with optical deformation larger than 11%, more A375MC2 cells (blue color pattern) will be selected than A375P (red color pattern).

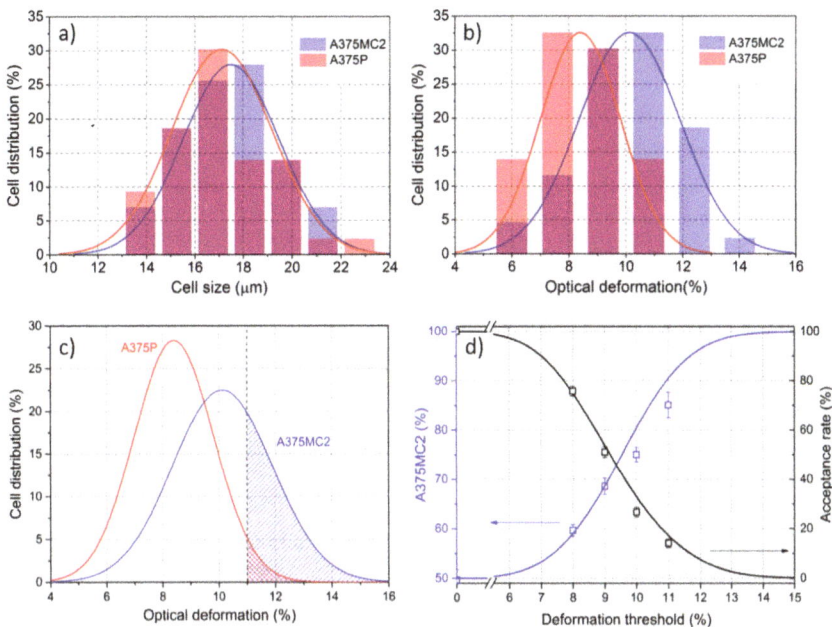

Figure 18. Cell sorting efficiency check. Characterization of the cellular size (**a**) and optical deformability (**b**) of two cell lines, A375MC2 and A375P; (**c**) Normalized cellular distributions as a function of their optical deformations from experiment data in (**b**). The whole area under each cell curve is set equal representing the same concentration. By defining a deformation threshold, a sub-population of A375MC2 can be enriched by collecting cells with higher deformability; (**d**) The ratio of A375MC2 in the collected cell sample and the ratio of cells in the initial sample that are expected to exhibit deformability higher than the threshold (acceptance rate) versus the defined threshold value. Figure reproduced from Reference [24] with permission from Royal Society of Chemistry.

In order to check the efficiency of this technique, it was necessary to have a method to calculate the percentage of A374P and A375MC2 cells in the collected cell sample, and this was achieved by pre-staining the A375MC2 cells with a fluorescent dye (LDS 751). By selecting different threshold value, this cell sorting experiment was repeated and in each collected cell sample, the percentage of A375MC was calculated by simply counting the ratio of fluorescent and non-fluorescent cells. The experimental results well matched with the theoretically expected values, but a deviation form the theoretical values was obtained when a high threshold value (e.g., 11%) was used. This is connected to the fact that by increasing the threshold, the acceptance rate (*i.e.*, the number of all collected cells divided by the number of all stretched cells) is reduced, thus making the measurement longer, and giving cells the possibility to deposit on the channel or to cluster, causing undesired perturbations in the system. Similar results were also observed in the study of Faigle *et al.* [25] suggesting that the active cell sorting based on cell mechanical properties can become a reliable and useful technique for the selection of specific sub-populations of very few cells.

6. Optical Heating and Temperature Effect

For laser application in cell biology, heating due to the absorption of optical radiation is an important issue that should be addressed. Indeed the possible thermal damage affect both the vitality of the samples and the validity of the results.

6.1. Optical Heating and Temperature Measurement

In an optical tweezer, tightly focused light beams are used, causing extremely high values of light intensity (in the order of magnitude of few MW/cm^2) to be produced in the beam focus, thus producing a significant increase of the local temperature. In particular, Peterman *et al.* [70] measured the temperature at the focus of the optical tweezer increased by 34.2 ± 0.1 K/W with 1064 nm laser for polystyrene beads of 2.2 μm diameter in glycerol medium. On the other side, OS is based on a completely different trapping configuration, obtained through two counter-propagating non-focused laser beams, as shown in Figure 1 and efficient cell trapping is achieved even using a low optical intensity from each fiber (on the order of a few mW over a large area). However, when cells-stretching measurements are conducted as described in Section 3.2, the sample flow is stopped, reducing the heat diffusion, and the trapping laser power is increased in the range between 0.5 and 1.5 W for each side, which can induce a non-negligible temperature increase.

In order to monitor the temperature change during cellular optical stretching, a precise measurement of the temperature value during the whole process should be performed. Moreover, the measurement should be done directly inside the microfluidic channel in the region illuminated by the laser radiation. However, the geometry of the micro-chip and its small dimension (100–300 μm for the internal channel) prevents the exploitation of conventional techniques as thermal sensor to perform the measurement. Additionally, it is difficult to have the spatially resolved temperature profile from the macroscopic sensor because it can only deliver an area-averaged results. All these issues were successfully overcome by a newly developed method called fluorescence ratio thermometry [71], in which laser-induced fluorescence (LIF) of two dyes, Rhodamine B and Rhodamine 110, is employed as a temperature indicator.

The first dye, Rhodamine B, has a well characterized temperature dependent LIF and also high temperature sensitivity (2.3% K^{-1}), while the second dye, Rhodamine 110, has temperature independent fluorescence and is therefore used as a reference. The temperature is dependent on the fluorescence intensity ratio of Rhodamine B and Rhodamine 110 and the combination usage of these two dyes can avoid the problems due to the local fluctuations from the excitation light intensity or from dye concentration. The other advantage of this method is that no normalization is required and absolute temperature can be directly measured with a resolution of 2 °C. The spatial temperature profile is obtained by measuring the fluorescence intensity ratio across the channel section at the central plane of the optical trap with a laser scanning confocal microscope with spatial resolution of less than 0.5 μm. Results of the temperature distribution are shown in Figure 19a,b for a total power of 2 W (1 W per side) in the trap, highlighting a temperature rise of 25 °C over a background temperature of 21 °C. It should be noted that the temperature spatial distribution is obtained after the temperature equilibrium is established and by averaging multiple successive scan. The same temperature measurements were repeated under different laser power values from 0.5 to 1.25 W per side and the results show a linear temperature dependence on the applied laser power, with a temperature increase rate of about 13 °C/W in the laser trap center.

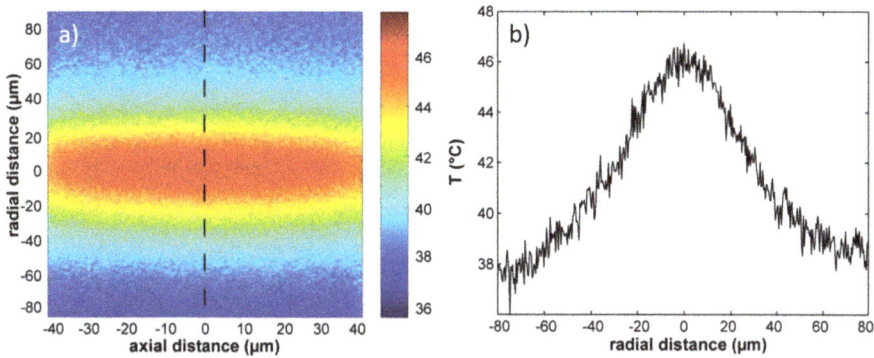

Figure 19. Spatial temperature profile in an optical stretcher. (**a**) Color image of the temperature increase from the two opposing laser beams in the optical stretcher. The imaging plane is the channel cross section through the center of the trap and the power of each laser beam is 1 W; (**b**) Line scan of the temperature along the dashed line in (**a**). Figure reproduced from Reference [71] with permission from Optical Society of America.

The same method was also applied to evaluate the temperature variation as a function of time [71–73], which is an important parameter, as during the stretching measurement the trapping power is abruptly increased for some seconds and is then lowered to the original trapping value, see Figure 20. In order to have a fast response and sufficient time resolution, the averaging procedure is removed and Figure 20 shows the temperature evolution for the step-like laser power increase. It can be seen that after the laser power is changed (either increased or decreased), the temperature equilibrium is simultaneously reached within fractions of a second. Wetzel *et al.* [72] also found the temperature immediately decreased to the equilibrium value after the laser is turned off.

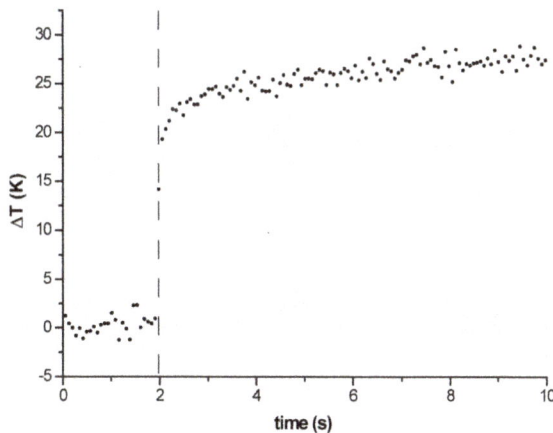

Figure 20. Temporal revolution of the temperature from the laser radiation in the optical stretcher. The laser is turned on at $t = 2$ s and has a total power of 2 W. Figure reproduced from Reference [71] with permission from Optical Society of America.

All the temperature measurements in space and time are performed when the microfluidic flux is stopped as maintaining a medium flow would easily remove the heat from the laser irradiated region. Additionally, it should be taken into account that the temperature increase from laser heating is strongly dependent on the laser light wavelength and the cell solution medium.

Another interesting effect produced by the presence of a temperature gradient is the creation of the so called Marangoni-effect, due to a surface-tension gradient because of the temperature profile. About this point we point out that the temperature gradient is non negligible only in the direction perpendicular to the beam axis, thus allowing us to neglect its impact on the elongation in the optical beams direction. Additionally, as the cell is trapped in the position corresponding to the maximum temperature, no net force is acting on the cell, and thus no displacement from the trapping position is observed.

6.2. Optical Heating Impact on Cell's Viability

As the OS is essentially used for single cell mechanics measurements, it is mandatory to investigate the optically induced heating impact on cell's viability and properties. In particular, it must be taken into account that the cytoskeleton may be altered by cell over-stretching as well as by heating, which would result into an unaccurate mechanical characterization. Differently from the conventional research done in this field, where small temperature increases (5–10 K) and longer time spans (minutes or hours) are applied [74], during optical stretching cells experience a high temperature increase in a short time duration as discussed in Section 6.1.

Standard proliferation and epigenetic analysis are the most accurate ways to verify cellular viability. However, they normally require large number of cells and several days to fulfill the assay, which makes it not applicable if the viability of each studied cell needs to be evaluated as soon as the measurement is performed. A simple approach has been proposed in the literature [8], which suggests to observe the cell appearance through phase contrast microscopy because dead cells usually show less contrast and no clear cellular contour. Despite its easiness, this method is not fully reliable being affected by man judgment. A more careful and accurate method is the use of the vital stain: Trypan Blue allows distinguishing viable cells because the dye can penetrate only the membrane of dead or damaged cells, which can not maintain their normal functionality, while it is excluded by viable ones. Even if this method allows for single cell evaluation, it can not be applied continuously after each cell stretching measurement because cells need to be removed from the microchannel for staining and checking. A method with on-site test capability is therefore needed and can be found in a different vital stain method with calcein acetoxymethylester (AM). This dye is membrane-permeable and, more importantly, is hydrolyzed by endogenous esterases into the green fluorescent calcein, which in turn only retains in the cytoplasm of living cells [75]. By measuring and comparing the green fluorescence intensity of each cell before and after optical stretching measurement, cell viability can thus be determined. However, photo-bleaching of 10%–20% from each time imaging extremely decreases its sensitivity.

A new method, even if with a very low throughput of less than 5 cells/h, based on cell spreading has been proposed specifically for this purpose and successfully demonstrated in literature [72]. After cell stretching measurement, the flux is not activated and the laser is completely turned off so as to let the studied cell slowly sink to the floor. After about 10 min, live cell could spread on the floor and attach to it. Since cell spreading is a vital feature of live cells, cell viability can be determined by observing cell spreading and attachment on the channel. This behavior of cell spreading involves the reassemble and rearrangement of the cell cytoskeleton and plasma and can be thus recognized as an effective indicator for cell viability. On the contrary, malignant cells show a very weak or absent spreading. By counting cells that show spreading ability, the cell viability can be easily obtained. Figure 21 shows the results of cell viability check after optical stretching in two different situations: in one case the temperature is increased by changing the laser power and in the other case by extending the stretching duration at the same laser power. It is found that more than 60% of cells can survive shorter laser heating of 0.5 s up to 58 ± 2 °C or can resist longer laser heating of 5 s at 48 ± 2 °C.

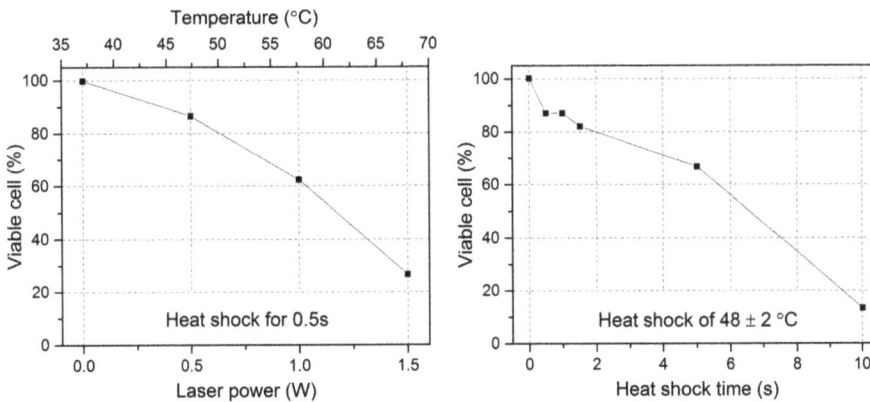

Figure 21. Heat shock impact on cell viability with different laser power (temperature) and time duration. The figure was realized exploiting the data reported in Reference [72].

6.3. Temperature Effect on Cell Mechanical Property

High laser powers are needed in order to induce an appreciable cell deformation during optical stretcher measurements. Although the beams are not focused and the cells are not directly damaged by the laser radiation, they might suffer from heating due to the high laser power. Moreover, in cell optical stretching measurement, the deformation could be produced by both optical force and optical heating. In order to evaluate their contributions and better understand the temperature effect on cell mechanical property, a series of studies have been presented [39,76–79], which include variations to the standard optical stretcher to monitor and modify the temperature on the cells.

In Figure 22 two examples are reported where the similar idea of realizing an active temperature control is shown. The first one exploits two additional fibers, positioned beside the optical stretching ones, to introduce additional laser heating, see Figure 22a, and the same wavelength of 1064 nm is coupled in all the fibers. Since the two heating fibers are very close to the active area of cell trapping and stretching, the temperature change can be built up within one second, hence a simple temperature control for optical stretching is obtained by simply changing the power injected in the two extra fibers. The second method obtains the same effect by coupling a second laser beam into one of the two stretching fibers, see Figure 22b. The wavelength of the heating laser is chosen as 1480 nm because its absorption is higher than that of 1064 nm wavelength, leading to a strong heating at low laser power, resulting in negligible optical force from the heating laser. By changing the heating laser power, temperature increase can be easily achieved. Both methods work locally inside the microfluidic circuit and they both allow only to increase the temperature. A different temperature control system that acts on the whole microchip is obtained by mounting it in a precisely regulated temperature environment, such as, a water bath [39], an aluminum sample holder [76,77] or a thermal chamber. Differently from the previously described methods, the thermal response in these cases is much slower and they are therefore intended for long scale temperature control.

These setups combined with an optical stretching device have been exploited to evaluate the temperature effect on cell mechanical property on different cell types, including human breast epithelial cell (MCF7, MCF10A), myeloid precursor cells (HL60) and human melanoma cells (A375MC2). Figure 23 reports an example obtained thorough the method based on extra heating-fibers, see Figure 22a. The optical compliance curves are obtained from the optical stretchering measurements both with different stretching laser power and without additional heating and with the same stretching power and different heating laser power. It can be observed that the stretching power increase leads to an strong decrease of cells' stiffness because of their stronger deformation. By applying the additional heating radiation, cells become softer at the same stretching laser power. A thermorheological methodology based on time-temperature superposition [76] has been exploited to explain these results and to propose that the cell creep behavior from optical stretching experiment is mainly due to the temperature effect from laser heating. Under long term temperature treatment and within a certain temperature threshold, cell stretching measurements still show similar result. Consequently, temperature increase induced by laser heating from the cell optical stretching process can have a strong effect on cell mechanical properties and should be always taken into account for a proper characterization.

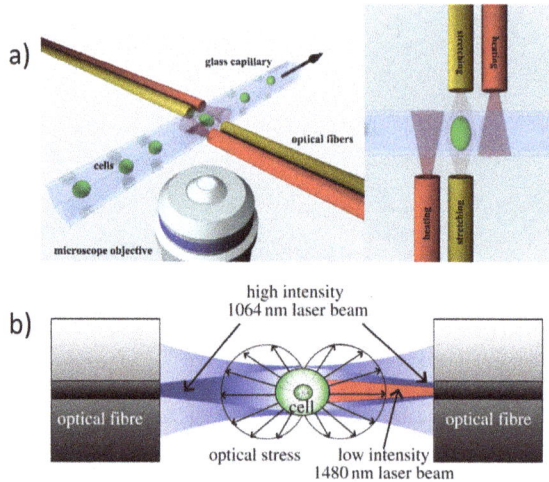

Figure 22. Active temperature control for optical stretcher. (**a**) Two additional fibers are added and positioned near the two opposing stretching fibers for temperature control; (**b**) Another laser is coupled into one of the two stretching fibers for temperature control. Figure reproduced from Reference [76,79] under CC-BY 3.0 license.

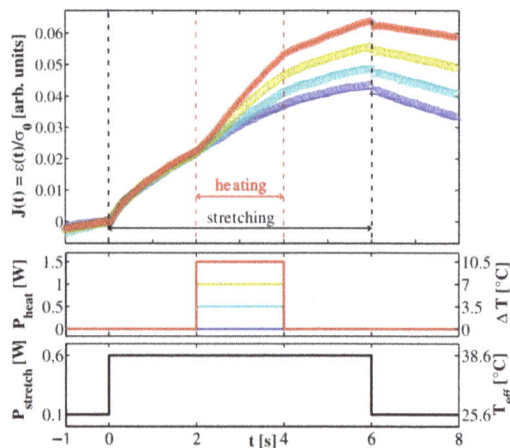

Figure 23. Temperature effect on cell optical deformation. Temperature is changed during the stretching measurement by using the two additional heating fibers (P_{heat}), see Figure 22a, while keeping the stretcher power ($P_{stretch}$) constant. Figure reproduced from Reference [76] under CC-BY 3.0 license.

7. Other Related Studies

Additionally to the above mentioned aspects and applications of the OS, new interesting studies are carried out by combining the OS with other techniques so as to integrate new functionalities. One example exploits resonant acoustic waves to prefocus the flowing cells at the correct channel height, so that they all intercept the laser beams and are therefore suitable for laser trapping and stretching. This idea was firstly proved by Khoury *et al.* [80] in a three layer assembled optical stretcher (a thin piece of PDMS with the microfluidic channel is sandwiched between two glass slides): the acoustic wave was driven by a piezo-ceramic attached beneath the chip. However, the plastic layer, having an acoustic impedance similar to that of water, allowed only for a low-efficiency excitation of the acoustic wave, and also required a fine selection of the bottom and top glass thickness.

A new study from Nava *et al.* recently exploited the same principle into an all-silica optical stretcher [37]. In Figure 24a the layout of the used monolithic OS is shown. Thanks to the use of a hard material, with an acoustic impedance very different from that of water, the efficiency of the acoustic wave excitation was strongly increased with respect to past result [80], and the use of a square-section channel allowed obtaining a 2D prefocusing effect (*i.e.*, acting both in vertical and horizontal direction) as shown in Figure 24b,c. The use of this chip allowed the users to greatly reduce the problems related to the height of flowing cells, thus doubling the OS measurement throughput. It was also demonstrated that the applied acoustic wave had no discernible effect on the cellular optical deformability on either red blood cells or mouse fibroblast cells. Besides, Yang *et al.* [81] further employed this chip in another recent study measuring both optical deformability and acoustic compressibility on single cells, by optical stretching and acoustophoresis experiments respectively. They found that the cancerous cell MDA-MB231 has both higher acoustic compressibility and higher optical deformability than its normal counterpart MCF7. And also, the optical deformability and acoustic compressibility are not correlated parameters. This result highlights the possibility to increase the functionalities in an optical stretcher to analyze cells in different aspects.

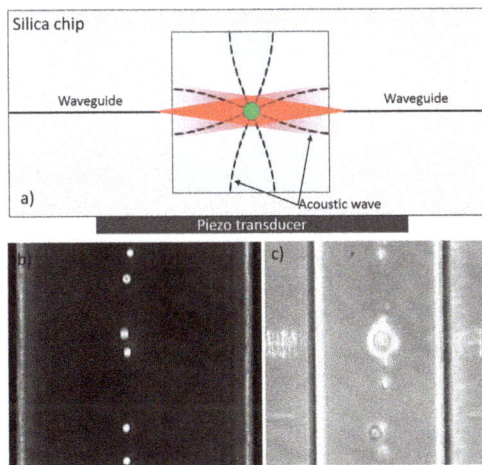

Figure 24. Acoustic prefocusing for optical stretcher. (**a**) Schematic illustration of the all glass microchip with both acoustic actuation (the black dash lines) driven by the underneath piezo ceramic and optical radiation (the red shaded area) emanating from the integrated waveguides. The microfluidic channel has a square cross section, 150 μm ×150 μm; (**b**) Microscope image of polystyrene beads trapped by acoustic wave in the middle of the microfluidic channel both horizontally and vertically (all beads are in the same focus); (**c**) Microscope image of red blood cells prefocused with acoustic wave for continuous optical stretching. Two opposing lasers from the waveguides are visible because of the light scattering.

Another example of the integration of new functionality in optical stretcher is the optical cell rotator. Differently from the optical trapping realized with two opposing single mode fibers in an assembled optical stretcher, Kreysing *et al.* [82] replaced one of the two fibers with a dual-mode fiber and spliced it with a defined offset with respect to the original single mode fiber. By rotating this dual mode fiber, the laser beam profile is changed accordingly, leading to an active rotation of the trapped cell. With this modification, they demonstrated that individual cell can be stably held with a well defined orientation or can be rotated perpendicularly to the laser axis. Recently, Kreysing *et al.* [83] even simplified this optical cell rotator by exploiting a few mode fiber and operating it dynamically beyond the single mode regime to realize precise optical field change without physically rotating the fiber itself (see. Figure 25). This ability to precisely orient cells in three dimensions could enable a range of applications in biological and medical research, such as the tomographic reconstruction of cell samples, by imaging them from different angles, the determination of the 3D refractive index distributions of live cells, or wide field fluorescent imaging from multiple angles with subsequent image fusion.

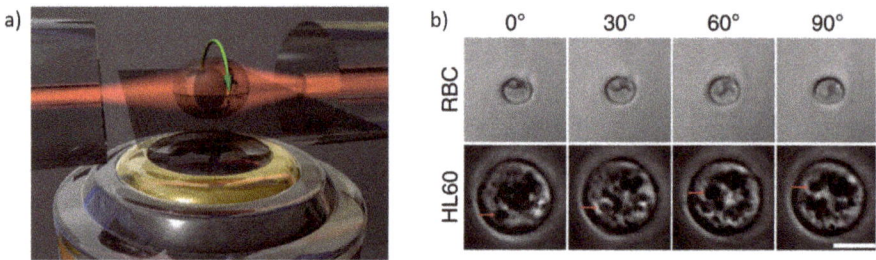

Figure 25. Optical cell rotator. (**a**) Illustration of the cell rotator realized in the two opposing laser radiation with an optical stretcher; (**b**) Microscope image sequences showing precise control of the cell orientation with red blood cell and HL60 cell. Figure reproduced from Reference [83] under CC-BY 4.0 license.

8. Final Discussion

Since the seminal works on optical forces by Arthur Ashkin, the possibility to carefully apply controlled forces to biological elements has attracted more and more attention. The development of the optical stretcher configuration, largely investigated by Guck, Kas *et al.*, created the basis for the optimization of a new, promising and flexible tool that can be used to trap, analyze and sort single cells. In this review we highlighted the most commonly exploited fabrication technologies, we described the physical effects at the basis of the optical stretcher working mechanism, and we offered a panoramic view on some of the most relevant applications.

Currently, the main limitation of the optical stretcher is its throughput, roughly about one cell per 10 s. Even the optical stretcher is already integrated within a microfluidic circuit and the optical deformation measurement can be realized automatically in real time, this throughput is still much lower than that of flow cytometry, which is based on purely hydrodynamic cell stretching and can offer a significantly high throughput of about 1000 cells per second. In this way, the future development of optical stretcher should be focused on the feature of the precise single cell characterization and, more importantly, the subsequent analysis on the same cell with other different functionalities. Two examples have been highlighted in Section 7, where the optical cell rotator for 3D imaging or the use of acoustic waves for prefocusing in continuous optical stretching measurements and compressibility measurement have been described. In addition, for optical stretcher fabrication, an all-polymer material technique can be possible as demonstrated by a study from Khoury *et al.* [84] showing that the polymer waveguides induced by deep UV lithography [85] are integrated with microfluidic channels and fully functional for optical manipulation.

We strongly believe that the development of microfluidic systems encompassing an optical-stretcher section will largely continue in the future, and will bring to the development of off-the-shelf devices for biologists and material-science researchers, especially thanks to the possibility to include optical stretching in microfluidic devices allowing for high-resolution imaging, 3D surface tomography and integrating multiple actuators systems.

Acknowledgments: The authors would like to thank the many co-workers involved in the activities reported in the present paper; in particular we acknowledge relevant contributions by Roberto Osellame from Istituto di Fotonica e Nanotecnologie - Consiglio Nazionale delle Ricerche, Chiara Mondello from Istituto di Genetica Molecolare - Consiglio Nazionale delle Ricerche, and Ilaria Cristiani from Università degli studi di Pavia.

Author Contributions: T.Y., F.B. and P.M. defined the manuscript organization and wrote the pape.

Conflicts of Interest: The authors declare no conflict of interest.

References

1. Suresh, S. Biomechanics and biophysics of cancer cells. *Acta Mater.* **2007**, *55*, 3989–4014.
2. Bao, G.; Suresh, S. Cell and molecular mechanics of biological materials. *Nat. Mater.* **2003**, *2*, 715–725.
3. Fletcher, D.A.; Mullins, R.D. Cell mechanics and the cytoskeleton. *Nature* **2010**, *463*, 485–492.
4. Moeendarbary, E.; Harris, A.R. Cell mechanics: Principles, practices, and prospects. *Wiley Interdiscip. Rev. Syst. Biol. Med.* **2014**, *6*, 371–388.
5. Kumar, S.; Weaver, V. Mechanics, malignancy, and metastasis: The force journey of a tumor cell. *Cancer Metastas. Rev.* **2009**, *28*, 113–127.
6. Denis, W.; Konstantinos, K.; Peter, C.S. The physics of cancer: The role of physical interactions and mechanical forces in metastasis. *Nat. Rev. Cancer* **2011**, *11*, 512–522.
7. Zhang, W.; Kai, K.; Choi, D.S.; Iwamoto, T.; Nguyen, Y.H.; Wong, H.; Landis, M.D.; Ueno, N.T.; Chang, J.; Qin, L. Microfluidics separation reveals the stem-cell-like deformability of tumor-initiating cells. *Proc. Natl. Acad. Sci. USA* **2012**, *109*, 18707–18712.
8. Guck, J.; Ananthakrishnan, R.; Mahmood, H.; Moon, T.J.; Cunningham, C.C.; Käs, J. The optical stretcher: A novel laser tool to micromanipulate cells. *Biophys. J.* **2001**, *81*, 767–784.
9. Ekpenyong, A.E.; Whyte, G.; Chalut, K.; Pagliara, S.; Lautenschläger, F.; Fiddler, C.; Paschke, S.; Keyser, U.F.; Chilvers, E.R.; Guck, J. Viscoelastic properties of differentiating blood cells are fate- and function-dependent. *PLoS ONE* **2012**, *7*, e45237.
10. Yap, B.; Kamm, R.D. Mechanical deformation of neutrophils into narrow channels induces pseudopod projection and changes in biomechanical properties. *J. Appl. Physiol.* **2005**, *98*, 1930–1939.
11. Vaziri, A.; Mofrad, M.R.K. Mechanics and deformation of the nucleus in micropipette aspiration experiment. *J. Biomech.* **2007**, *40*, 2053–2062.
12. Pachenari, M.; Seyedpour, S.M.; Janmaleki, M.; Babazadeh Shayan, S.; Taranejoo, S.; Hosseinkhani, H. Mechanical properties of cancer cytoskeleton depend on actin filaments to microtubules content: Investigating different grades of colon cancer cell lines. *J. Biomech.* **2014**, *47*, 373–379.
13. Mathur, A.B.; Collinsworth, A.M.; Reichert, W.M.; Kraus, W.E.; Truskey, G.A. Endothelial, cardiac muscle and skeletal muscle exhibit different viscous and elastic properties as determined by atomic force microscopy. *J. Biomech.* **2001**, *34*, 1545–1553.
14. Mackay, J.L.; Kumar, S. Measuring the elastic properties of living cells with atomic force microscopy indentation. *Methods Mol. Biol.* **2013**, *931*, 313–329.
15. Rouven Brückner, B.; Pietuch, A.; Nehls, S.; Rother, J.; Janshoff, A. Ezrin is a major regulator of membrane tension in epithelial cells. *Sci. Rep.* **2015**, *5*, 14700.
16. Dao, M.; Lim, C.; Suresh, S. Mechanics of the human red blood cell deformed by optical tweezers. *J. Mech. Phys. Solids* **2003**, *51*, 2259–2280.
17. Chen, J.; Fabry, B.; Schiffrin, E.L.; Wang, N. Twisting integrin receptors increases endothelin-1 gene expression in endothelial cells. *Am. J. Physiol. Cell Physiol.* **2001**, *280*, C1475–C1484.
18. Preira, P.; Valignat, M.P.; Bico, J.; Théodoly, O. Single cell rheometry with a microfluidic constriction: Quantitative control of friction and fluid leaks between cell and channel walls. *Biomicrofluidics* **2013**, *7*, 024111.

19. Luo, Y.N.; Chen, D.Y.; Zhao, Y.; Wei, C.; Zhao, X.T.; Yue, W.T.; Long, R.; Wang, J.B.; Chen, J. A Constriction channel based microfluidic system enabling continuous characterization of cellular instantaneous young's modulus. *Sens. Actuators B Chem.* **2014**, *202*, 1183–1189.

20. Martinez Vazquez, R.; Nava, G.; Veglione, M.; Yang, T.; Bragheri, F.; Minzioni, P.; Bianchi, E.; Di Tano, M.; Chiodi, I.; Osellame, R.; *et al.* An optofluidic constriction chip for monitoring metastatic potential and drug response of cancer cells. *Integr. Biol.* **2015**, *7*, 477–484.

21. Otto, O.; Rosendahl, P.; Mietke, A.; Golfier, S.; Herold, C.; Klaue, D.; Girardo, S.; Pagliara, S.; Ekpenyong, A.; Jacobi, A.; *et al.* Real-time deformability cytometry: On-the-fly cell mechanical phenotyping. *Nat. Methods* **2015**, *12*, 199–202.

22. Mietke, A.; Otto, O.; Girardo, S.; Rosendahl, P.; Taubenberger, A.; Golfier, S.; Ulbricht, E.; Aland, S.; Guck, J.; Fischer-Friedrich, E. Extracting cell stiffness from real-time deformability cytometry: Theory and experiment. *Biophys. J.* **2015**, *109*, 2023–2036.

23. Guck, J.; Schinkinger, S.; Lincoln, B.; Wottawah, F.; Ebert, S.; Romeyke, M.; Lenz, D.; Erickson, H.M.; Ananthakrishnan, R.; Mitchell, D.; *et al.* Optical deformability as an inherent cell marker for testing malignant transformation and metastatic competence. *Biophys. J.* **2005**, *88*, 3689–3698.

24. Yang, T.; Paiè, P.; Nava, G.; Bragheri, F.; Vazquez, R.M.; Minzioni, P.; Veglione, M.; Di Tano, M.; Mondello, C.; Osellame, R.; *et al.* An integrated optofluidic device for single-cell sorting driven by mechanical properties. *Lab Chip* **2015**, *15*, 1262–1266.

25. Faigle, C.; Lautenschläger, F.; Whyte, G.; Homewood, P.; Martín-Badosa, E.; Guck, J. A monolithic glass chip for active single-cell sorting based on mechanical phenotyping. *Lab Chip* **2015**, *15*, 1267–1275.

26. Lincoln, B.; Schinkinger, S.; Travis, K.; Wottawah, F.; Ebert, S.; Sauer, F.; Guck, J. Reconfigurable microfluidic integration of a dual-beam laser trap with biomedical applications. *Biomed. Microdevices* **2007**, *9*, 703–710.

27. Ren, K.; Zhou, J.; Wu, H. Materials for microfluidic chip fabrication. *Acc. Chem. Res.* **2013**, *46*, 2396–2406.

28. Guck, J.; Ananthakrishnan, R.; Moon, T.J.; Cunningham, C.C.; Käs, J. Optical deformability of soft biological dielectrics. *Phys. Rev. Lett.* **2000**, *84*, 5451–5454.

29. Bellini, N.; Bragheri, F.; Cristiani, I.; Guck, J.; Osellame, R.; Whyte, G. Validation and perspectives of a femtosecond laser fabricated monolithic optical stretcher. *Biomed. Opt. Express* **2012**, *3*, 2658–2668.

30. Matteucci, M.; Triches, M.; Nava, G.; Kristensen, A.; Pollard, M.; Berg-Sørensen, K.; Taboryski, R. Fiber-based, injection-molded optofluidic systems: Improvements in assembly and applications. *Micromachines* **2015**, *6*, 1971–1983.

31. Gattass, R.R.; Mazur, E. Femtosecond laser micromachining in transparent materials. *Nat. Photonics* **2008**, *2*, 219–225.

32. He, F.; Liao, Y.; Lin, J.; Song, J.; Qiao, L.; Cheng, Y.; Sugioka, K. Femtosecond laser fabrication of monolithically integrated microfluidic sensors in glass. *Sensors* **2014**, *14*, 19402–19440.

33. Osellame, R.; Maselli, V.; Vazquez, R.M.; Ramponi, R.; Cerullo, G. Integration of optical waveguides and microfluidic channels both fabricated by femtosecond laser irradiation. *Appl. Phys. Lett.* **2007**, *90*, 231118.

34. Osellame, R.; Hoekstra, H.; Cerullo, G.; Pollnau, M. Femtosecond laser microstructuring: An enabling tool for optofluidic lab-on-chips. *Laser Photonics Rev.* **2011**, *5*, 442–463.

35. Taylor, R.S.; Hnatovsky, C.; Simova, E.; Rayner, D.M.; Bhardwaj, V.R.; Corkum, P.B. Femtosecond laser fabrication of nanostructures in silica glass. *Opt. Lett.* **2003**, *28*, 1043–1045.

36. Bellini, N.; Vishnubhatla, K.C.; Bragheri, F.; Ferrara, L.; Minzioni, P.; Ramponi, R.; Cristiani, I.; Osellame, R. Femtosecond laser fabricated monolithic chip for optical trapping and stretching of single cells. *Opt. Express* **2010**, *18*, 4679–4688.

37. Nava, G.; Bragheri, F.; Yang, T.; Minzioni, P.; Osellame, R.; Cristiani, I.; Berg-Sørensen, K. All-silica microfluidic optical stretcher with acoustophoretic prefocusing. *Microfluid. Nanofluid.* **2015**, *19*, 837–844.

38. Bragheri, F.; Ferrara, L.; Bellini, N.; Vishnubhatla, K.C.; Minzioni, P.; Ramponi, R.; Osellame, R.; Cristiani, I. Optofluidic chip for single cell trapping and stretching fabricated by a femtosecond laser. *J. Biophotonics* **2010**, *3*, 234–243.

39. Yang, T.; Nava, G.; Minzioni, P.; Veglione, M.; Bragheri, F.; Lelii, F.D.; Vazquez, R.M.; Osellame, R.; Cristiani, I. Investigation of temperature effect on cell mechanics by optofluidic microchips. *Biomed. Opt. Express* **2015**, *6*, 2991–2996.

40. Ashkin, A. Forces of a single-beam gradient laser trap on a dielectric sphere in the ray optics regime. *Biophys. J.* **1992**, *61*, 569–582.

41. Ashkin, A. Trapping of atoms by resonance radiation pressure. *Phys. Rev. Lett.* **1978**, *40*, 729–732.
42. Ashkin, A.; Dziedzic, J.M. Radiation pressure on a free liquid surface. *Phys. Rev. Lett.* **1973**, *30*, 139–142.
43. Ferrara, L.; Baldini, E.; Minzioni, P.; Bragheri, F.; Liberale, C.; Fabrizio, E.D.; Cristiani, I. Experimental study of the optical forces exerted by a Gaussian beam within the Rayleigh range. *J. Opt.* **2011**, *13*, 75712–75718.
44. Bareil, P.B.; Sheng, Y.; Chiou, A. Local scattering stress distribution on surface of a spherical cell in optical stretcher. *Opt. Express* **2006**, *14*, 12503–12509.
45. Bareil, P.B.; Sheng, Y.; Chen, Y.Q.; Chiou, A. Calculation of spherical red blood cell deformation in a dual-beam optical stretcher. *Opt. Express* **2007**, *15*, 16029–16034.
46. Ekpenyong, A.E.; Posey, C.L.; Chaput, J.L.; Burkart, A.K.; Marquardt, M.M.; Smith, T.J.; Nichols, M.G. Determination of cell elasticity through hybrid ray optics and continuum mechanics modeling of cell deformation in the optical stretcher. *Appl. Opt.* **2009**, *48*, 6344–6354.
47. Brevik, I. Experiments in phenomenological electrodynamics and the electromagnetic energy-momentum tensor. *Phys. Rep.* **1979**, *52*, 133–201.
48. Lincoln, B.; Wottawah, F.; Schinkinger, S.; Ebert, S.; Guck, J. High throughput rheological measurements with an optical stretcher. *Methods Cell Biol.* **2007**, *83* 397–423.
49. Wottawah, F.; Schinkinger, S.; Lincoln, B.; Ananthakrishnan, R.; Romeyke, M.; Guck, J.; Käs, J. Optical rheology of biological cells. *Phys. Rev. Lett.* **2005**, *94*, 98103.
50. Wottawah, F.; Schinkinger, S.; Lincoln, B.; Ebert, S.; Müller, K.; Sauer, F.; Travis, K.; Guck, J. Characterizing single suspended cells by optorheology. *Acta Biomater.* **2005**, *1*, 263–271.
51. Ananthakrishnan, R.; Guck, J.; Wottawah, F.; Schinkinger, S.; Lincoln, B.; Romeyke, M.; Kas, J. Modelling the structural response of an eukaryotic cell in the optical stretcher. *Curr. Sci.* **2005**, *88*, 1434–1440.
52. Ananthakrishnan, R.; Guck, J.; Wottawah, F.; Schinkinger, S.; Lincoln, B.; Romeyke, M.; Moon, T.; Käs, J. Quantifying the contribution of actin networks to the elastic strength of fibroblasts. *J. Theor. Biol.* **2006**, *242*, 502–516.
53. Gladilin, E.; Gonzalez, P.; Eils, R. Dissecting the contribution of actin and vimentin intermediate filaments to mechanical phenotype of suspended cells using high-throughput deformability measurements and computational modeling. *J. Biomech.* **2014**, *47*, 2598–2605.
54. Ye, T.; Phan-Thien, N.; Khoo, B.C.; Lim, C.T. Stretching and relaxation of malaria-infected red blood cells. *Biophys. J.* **2013**, *105*, 1103–1109.
55. Mauritz, J.A.; Esposito, A.; Tiffert, T.; Skepper, J.; Warley, A.; Yoon, Y.Z.; Cicuta, P.; Lew, V.; Guck, J.; Kaminski, C. Biophotonic techniques for the study of malaria-infected red blood cells. *Med. Biol. Eng. Comput.* **2010**, *48*, 1055–1063.
56. Sawetzki, T.; Eggleton, C.D.; Desai, S.A.; Marr, D.W.M. Viscoelasticity as a biomarker for high-throughput flow cytometry. *Biophys. J.* **2013**, *105*, 2281–2288.
57. Sraj, I.; Eggleton, C.D.; Jimenez, R.; Hoover, E.; Squier, J.; Chichester, J.; Marr, D.W.M. Cell deformation cytometry using diode-bar optical stretchers. *J. Biomed. Opt.* **2010**, *15*, 47010.
58. Sraj, I.; Marr, D.W.M.; Eggleton, C.D. Linear diode laser bar optical stretchers for cell deformation. *Biomed. Opt. Express* **2010**, *1*, 482–488.
59. Liu, Y.P.; Liu, K.K.; Lai, A.C.K.; Li, C. The deformation of an erythrocyte under the radiation pressure by optical stretch. *J. Biomech. Eng.* **2006**, *128*, 830–836.
60. Tan, Y.; Sun, D.; Huang, W. Mechanical modeling of red blood cells during optical stretching. *J. Biomech. Eng.* **2010**, *132*, 044504.
61. Solmaz, M.E.; Biswas, R.; Sankhagowit, S.; Thompson, J.R.; Mejia, C.A.; Malmstadt, N.; Povinelli, M.L. Optical stretching of giant unilamellar vesicles with an integrated dual-beam optical trap. *Biomed. Opt. Express* **2012**, *3*, 2419–2427.
62. Delabre, U.; Feld, K.; Crespo, E.; Whyte, G.; Sykes, C.; Seifert, U.; Guck, J. Deformation of phospholipid vesicles in an optical stretcher. *Soft Matter* **2015**, *11*, 6075–6088.
63. Lautenschläger, F.; Paschke, S.; Schinkinger, S.; Bruel, A.; Beil, M.; Guck, J. The regulatory role of cell mechanics for migration of differentiating myeloid cells. *Proc. Natl. Acad. Sci. USA* **2009**, *106*, 15696–15701.
64. Schulze, C.; Wetzel, F.; Kueper, T.; Malsen, A.; Muhr, G.; Jaspers, S.; Blatt, T.; Wittern, K.P.; Wenck, H.; Käs, J.A. Stiffening of human skin fibroblasts with age. *Clin. Plast. Surg.* **2012**, *39*, 9–20.
65. Bragheri, F.; Minzioni, P.; Martinez Vazquez, R.; Bellini, N.; Paie, P.; Mondello, C.; Ramponi, R.; Cristiani, I.; Osellame, R. Optofluidic integrated cell sorter fabricated by femtosecond lasers. *Lab Chip* **2012**, *12*, 3779–3784.

66. Holmes, D.; Whyte, G.; Bailey, J.; Vergara-Irigaray, N.; Ekpenyong, A.; Guck, J.; Duke, T. Separation of blood cells with differing deformability using deterministic lateral displacement. *Interface Focus* **2014**, *4*, UNSP 20140011.
67. Guo, Q.; Duffy, S.P.; Matthews, K.; Deng, X.; Santoso, A.T.; Islamzada, E.; Ma, H. Deformability based sorting of red blood cells improves diagnostic sensitivity for malaria caused by Plasmodium falciparum. *Lab Chip* **2016**, *16*, 645–654.
68. Beech, J.P.; Holm, S.H.; Adolfsson, K.; Tegenfeldt, J.O. Sorting cells by size, shape and deformability. *Lab Chip* **2012**, *12*, 1048–1051.
69. Wang, G.; Mao, W.; Byler, R.; Patel, K.; Henegar, C.; Alexeev, A.; Sulchek, T. Stiffness dependent separation of cells in a microfluidic device. *PLoS ONE* **2013**, *8*, e75901.
70. Peterman, E.J.G.; Gittes, F.; Schmidt, C.F. Laser-induced heating in optical traps. *Biophys. J.* **2003**, *84*, 1308–1316.
71. Ebert, S.; Travis, K.; Lincoln, B.; Guck, J. Fluorescence ratio thermometry in a microfluidic dual-beam laser trap. *Opt. Express* **2007**, *15*, 15493–15499.
72. Wetzel, F.; Rönicke, S.; Müller, K.; Gyger, M.; Rose, D.; Zink, M.; Käs, J. Single cell viability and impact of heating by laser absorption. *Eur. Biophys. J. EBJ* **2011**, *40*, 1109–1114.
73. Gyger, M.; Rose, D.; Stange, R.; Kiessling, T.; Zink, M.; Fabry, B.; Käs, J.A. Calcium imaging in the optical stretcher. *Opt. Express* **2011**, *19*, 19212–19222.
74. Huang, S.H.; Yang, K.J.; Wu, J.C.; Chang, K.J.; Wang, S.M. Effects of hyperthermia on the cytoskeleton and focal adhesion proteins in a human thyroid carcinoma cell line. *J. Cell. Biochem.* **1999**, *75*, 327–337.
75. Wang, X.M.; Terasaki, P.I.; Rankin, G.W.; Chia, D.; Zhong, H.P.; Hardy, S. A new microcellular cytotoxicity test based on calcein AM release. *Hum. Immunol.* **1993**, *37*, 264–270.
76. Kießling, T.R.; Stange, R.; Käs, J.A.; Fritsch, A.W. Thermorheology of living cells-impact of temperature variations on cell mechanics. *New J. Phys.* **2013**, *15*, 045026.
77. Warmt, E.; Kießling, T.R.; Stange, R.; Fritsch, A.W.; Zink, M.; Käs, J.A. Thermal instability of cell nuclei. *New J. Phys.* **2014**, *16*, 073009.
78. Schmidt, B.U.S.; Kießling, T.R.; Warmt, E.; Fritsch, A.W.; Stange, R.; Käs, J.A. Complex thermorheology of living cells. *New J. Phys.* **2015**, *17*, 073010.
79. Chan, C.J.; Whyte, G.; Boyde, L.; Salbreux, G.; Guck, J. Impact of heating on passive and active biomechanics of suspended cells. *Interface Focus* **2014**, *4*, 20130069.
80. Khoury, M.; Barnkob, R.; Laub Busk, L.; Tidemand-Lichtenberg, P.; Bruus, H.; Berg-Sørensen, K. Optical stretching on chip with acoustophoretic prefocusing. *Proc. SPIE* **2012**, *8458*, 84581E.
81. Yang, T.; Bragheri, F.; Nava, G.; Chiodi, I.; Mondello, C.; Osellame, R.; Berg-Sørensen, K.; Cristiani, I.; Minzioni, P. A comprehensive strategy for the analysis of acoustic compressibility and optical deformability on single cells. *Sci. Rep.* **2016**, *6*, 23946.
82. Kreysing, M.K.; Kießling, T.; Fritsch, A.; Dietrich, C.; Guck, J.R.; Käs, J.A. The optical cell rotator. *Opt. Express* **2008**, *16*, 16984–16992.
83. Kreysing, M.; Ott, D.; Schmidberger, M.J.; Otto, O.; Schürmann, M.; Martín-Badosa, E.; Whyte, G.; Guck, J. Dynamic operation of optical fibres beyond the single-mode regime facilitates the orientation of biological cells. *Nat. Commun.* **2014**, *5*, 5481.
84. Khoury, M.; Vannahme, C.; Sørensen, K.; Kristensen, A.; Berg-Sørensen, K. Monolithic integration of DUV-induced waveguides into plastic microfluidic chip for optical manipulation. *Microelectron. Eng.* **2014**, *121*, 5–9.
85. Tomlinson, W.J.; Kaminow, I.P.; Chandross, E.A.; Fork, R.L.; Silfvast, W.T. Photo induced refractive index increase in poly(Methylmethacrylate) and its application. *Appl. Phys. Lett.* **1970**, *16*, 486–489.

micromachines

MDPI

Review

Optofluidic Device Based Microflow Cytometers for Particle/Cell Detection: A Review

Yushan Zhang [1], Benjamin R. Watts [2], Tianyi Guo [1], Zhiyi Zhang [3], Changqing Xu [4,*] and Qiyin Fang [4]

[1]   School of Biomedical Engineering, McMaster University, 1280 Main Street West, Hamilton, ON L8S 4L8, Canada; zhang749@mcmaster.ca (Y.Z.); guot2@mcmaster.ca (T.G.)

[2]   ArtIC Photonics, 260 Terence Matthews Cres, Ottawa, ON K2M 2C7, Canada; benjamin.r.watts@gmail.com

[3]   Information and Communication Technologies, National Research Council of Canada, 1200 Montreal Road, Ottawa, ON K1A 0R6, Canada; zhiyi.zhang@nrc-cnrc.gc.ca

[4]   Department of Engineering Physics, McMaster University, 1280 Main Street West, Hamilton, ON L8S 4L8, Canada; qiyin.fang@mcmaster.ca

*   Correspondence: cqxu@mcmaster.ca; Tel.: +1-905-525-9140 (ext. 24314)

Academic Editors: Shih-Kang Fan and Nam-Trung Nguyen
Received: 1 March 2016; Accepted: 12 April 2016; Published: 15 April 2016

Abstract: Optofluidic devices combining micro-optical and microfluidic components bring a host of new advantages to conventional microfluidic devices. Aspects, such as optical beam shaping, can be integrated on-chip and provide high-sensitivity and built-in optical alignment. Optofluidic microflow cytometers have been demonstrated in applications, such as point-of-care diagnostics, cellular immunophenotyping, rare cell analysis, genomics and analytical chemistry. Flow control, light guiding and collecting, data collection and data analysis are the four main techniques attributed to the performance of the optofluidic microflow cytometer. Each of the four areas is discussed in detail to show the basic principles and recent developments. 3D microfabrication techniques are discussed in their use to make these novel microfluidic devices, and the integration of the whole system takes advantage of the miniaturization of each sub-system. The combination of these different techniques is a spur to the development of microflow cytometers, and results show the performance of many types of microflow cytometers developed recently.

Keywords: optofluidic device; microfluidics; microflow cytometer; microfabrication

1. Introduction

Since the original attempt in 1934 when researchers first successfully counted particles and cells in a small tube [1], flow cytometry has developed into a powerful technique for cell analysis, sorting and counting. Recently, flow cytometry has been applied in many fields, such as point-of-care (POC) diagnostics, cellular immunophenotyping, rare cell analysis and genomics [2]. The commercialization of conventional flow cytometers has been very successful: the market of modern microflow cytometers is expected to reach \$3.6–5.7 billion by 2018 at a compound annual growth rate of 18%–29% [3]. Compared to a bulky conventional flow cytometer, microchip-based flow cytometers (referred to as microflow cytometers in this paper) are simple to use, time efficient, consume low amounts of expensive reagents and have overall lower associated costs (capital, operation, training, *etc*). With the rapidly developing demands of POC applications, the growing demands of *in situ* and *in vitro* diagnostics in the biomedical field and the need to improve rapid analysis and synthesis in the chemical field, microflow cytometers are poised to facilitate great advancement in these and other fields and allow applications that will change many aspects of everyday life in the near future.

The term "optofluidics" was first mentioned in 2003 and was coined to reference new devices that integrated the fields of optics and microfluidics [4]. Microfluidics is the technology that manipulates fluids on the nL–fL scale on a microchip platform, whereas optofluidics manipulates both fluids and optics simultaneously in a seamlessly integrated platform. A microchip-based device that is based on the technology of microfluidics is called a microfluidic device, while an optofluidic device is a device based on optofluidics, requiring both fluidic and optical capabilities. A microflow cytometer is a highly integrated system that utilizes a microchip-based device for the fluidic handling and manipulation in a flow cytometry application. An optofluidic microflow cytometer utilizes an optofluidic device to apply flow cytometry using a single device to integrate both the fluidic and optical sub-systems onto a single device. Classification of the terminology and the function of the device is shown in Table 1.

In an optofluidic microflow cytometer, the optical components are integrated into a microfluidic system and *vice versa* [5]. Integration allows the benefits of including new optical features on the device, such as built-in optical alignment, beam shaping, high optical sensitivity and tenability-each seamlessly integrated in one platform with fluidics. Previous review papers have already summarized the fundamentals and applications of optofluidic technology [4–6]. In this paper, we will discuss the basic principles and components of an optofluidic device-based microflow cytometer in detail, as well as review the performance of a few microflow cytometers developed recently and compare the performance of the devices.

Table 1. Terminology.

Terminology	Main Device Used	Description
Microflow cytometer	Microfluidic device	Integrated optics are not necessary
Optofluidic microflow cytometer	Optofluidic device	Integrated optics are necessary

1.1. The Principles of Flow Cytometry

Flow cytometry is a powerful analysis technology for the characterization of cells or particles. Multiple parameters, *i.e.*, size, shape, cell granularity and cell viability, can be detected simultaneously at rates of up to 50,000 particles per second [1]. The original intention of flow cytometry was to measure particles or cells one-by-one as they passed through a laser beam in a single file stream flowing through a glass tube [7]. Scattered light at both small and large angles, as well as fluorescence light (FL) emitted from fluorescent labels are detected and analyzed for every single cell or particle. Light scattered at a small angle from the input beam axis is referred to as forward scattered light (FSC), whereas large angles of scattered light are called side scattered light (SSC). The intensity of FSC is generally determined by the size of the cell, while the granularity of the particle or cell determines the intensity of SSC. Through the analysis of the two data parameters, the cells or particles can be identified, counted and sorted downstream. The performance of a flow cytometer is dependent on one of the four main techniques integral to flow cytometry: particle focusing, beam shaping, signal detection and data analysis.

In flow cytometry, particle focusing is applied to ensure the cells of interest pass through the optical interrogation point one by one, reducing the possibility of a double detection. The interrogation region is the intersection between the excitation light beam and the solid angle accepting scattered light or fluorescence from the detection optics. The sample fluid containing cells or particles is surrounded by a sheath fluid which confines the particles to a narrow stream in the center of the channel, roughly one cell or particle in diameter.

Conventional flow cytometers focus and shape the excitation beam by using a free-space lens system. Ideally, the beam would be aligned with focused sample stream, and the beam width would be no less than the width of sample stream to ensure the entire particle or cell can be illuminated. In addition, the portion of the light beam outside the sample stream is minimized to keep the background illumination, thus the noise on the detection channels, as low as possible.

A pulse is produced when a particle or cell passes through the laser beam. The pulse shape and amplitude relates to the interaction between the incident beam and the particle or cell, including the incident light intensity, particle size, geometry and granularity, as well as the fluorescence efficiency. The pulse duration depends on the beam width and linear velocity of the particle along the channel. Therefore, light beam intensity with a super-Gaussian distribution along the flow direction is preferred to generate pulses close to a square waveform.

When particles pass through the laser beam, SSC, FSC and fluorescence light signals are detected by a number of detectors. A modern flow cytometer can detect as many as 17 independent channels featuring a combination of several FL wavelengths, FSC and several different angles of SSC simultaneously by using a series of dichroic mirrors [1]. Thus, multi-parameters can be monitored by analyzing those light signals. In some circumstances, in conjunction with or in place of an optical interrogation method, impedance-based cell or a particle sorting, counting and differentiating method can also be applied in flow cytometry [8,9]. Cells or particles pass through a small area enclosed by two electrodes, where electrophysiological impedance variation can be detected for every single cell or particle.

1.2. Microflow Cytometer

Although conventional flow cytometers have gained profound success in cell sorting and analysis, they are bulky, demand large amounts of expensive reagents with complicated processing steps, are complicated in manipulation and require high maintenance costs. Typically, a flow cytometer will be located in a single facility where many hundreds of users will need access to it. These limitations restrict their uses in POC diagnostics, *in situ* pathogen monitoring and other application where portability, handling small volume samples, low operation costs and ease of operation are essential.

Owing to the recent development of lab-on-chip (LOC) technology, microfabrication and micromachining techniques, the miniaturization of a flow cytometer can be achieved. The microfabrication of a flow cytometer with 3D microstructures can be accomplished by 3D microfabrication techniques utilizing UV lithography [10]. Microfluidic mixing [11], microfluidic cell sorting and the miniaturization of pumps and valves [12] provide a basic foundation for miniaturizing the fluidic handling to develop a microchip-based flow cytometer. Researchers have been able to take advantage of these advanced microfabrication technologies to miniaturize a flow cytometer to a microscale or even a nanoscale platform. Controlling fluids in a microchannel allows microchip-based flow cytometer to be applied in POC diagnostics and lab-on-a-chip devices offers unique advantages [13], such as reducing the volumes of reagents, shortening the turnaround time between inspection and results and lowering the associated costs [14].

To date, the throughput of microflow cytometers can reach up to 50,000 cells/s [1], allowing microflow cytometers to have many applications: Titmarsh *et al.* [15] discussed how microfluidic technology spurred on the development of stem cell-derived therapies. Hashemi *et al.* [16] successfully distinguished different populations of phytoplankton with high sensitivity by measuring light scatter and fluorescence properties by a microflow cytometer. More demonstrations on diagnostic and point-of-care applications have been addressed in recent review papers [14,17].

A novel optofluidic device-based microflow cytometer emerged recently. Optical components and novel liquid lenses are used in microfluidic devices. Liquid-core/liquid-cladding waveguides and liquid core/air-cladding lens systems with larger refractive index contrast lead to less propagation losses and resulted in better optical confinement [18]. Additionally, an integrated on-chip lens system or grooved on-chip fibers further reduce the size of the microflow cytometer. Built-in waveguides also are free of optical alignment, making operation much easier. Recently, Liang *et al.* took advantage of evanescent waves present at the liquid-liquid interface of immiscible flows to count the nanoparticles on an optofluidic microchip [19].

2. Major Components of an Optofluidic Microflow Cytometer

Typically, the creation of an optofluidic device-based microflow cytometer includes four principle design areas: (1) the flow control; (2) the optical design; (3) the microfabrication of functional layers; (4) the integration of the entire system. The flow control includes how to bring flow into a device and to ensure cells or particles are being focused in the interrogation region, which is usually achieved by 2D or 3D hydrodynamic focusing methods. The optical system provides light for interaction and collects light signals for analysis. The microfabrication of fluid control and optical components provide a microscale or nanoscale platform for cell analysis. System integration includes the miniaturization of the device and provides user-friendly control environment and easy-to-use data analysis software.

Figure 1 shows a system setup of a typical optofluidic microflow cytometer [20]. Cells are delivered to the interrogation region in a sample fluid surrounded by two sheath fluids. Cells traverse the light in the interrogation region that has been focused by the on-chip lens system and produces its characteristic optical signature containing SSC, FSC and FL signals. In this iteration of the device, the collection arm is not integrated on the device like the excitation optics, and thus, the signals are collected via a free space objective and directed to a spectral and spatial filter where they are finally detected and amplified by a photomultiplier tube (PMT). In this device, the bulky and expensive free space optical lens system for excitation in a conventional flow cytometer was replaced by cost-effective, space-saving and free optical alignment on-chip lens system.

The challenge and difficulty of miniaturizing the flow cytometer is how to make the performance of a microflow cytometer comparable to the conventional benchtop flow cytometer. In this chapter, optofluidic microflow cytometers with different features classified in Table 2 will be discussed. All of the aspects in Table 2 contribute to the performance of an optofluidic microflow cytometer and will be discussed in detail in later sections.

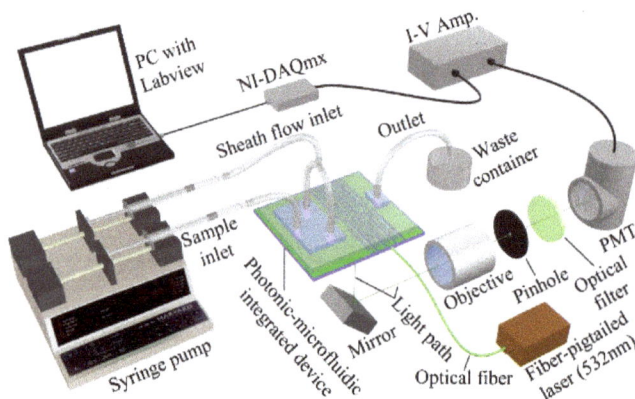

Figure 1. System setup of an optofluidic device (also referred to as photonic-microfluidic integrated device)-based microflow cytometer. Laser light is focused through on-chip lenses, and side scattered light (SSC) and fluorescence light (FL) signals are detected via free-space lens system. Light signals are amplified by a photomultiplier tube (PMT), then data are analyzed by a data acquisition card (DAQ). Reproduced from [20] with kind permission from Wiley.

Table 2. Techniques related to the performance of an optofluidic microflow cytometer.

Flow Control	Light Guide	Light Collection	Collected Signal
2D hydrodynamic focusing	Free-space/on-chip	Free-space/on-chip	Fluorescence collection (FL)
3D hydrodynamic focusing	Free-space/on-chip	Free-space/on-chip	Side scattered light (SSC)
Other methods	Free-space	Free-space	Forward scattered light (FSC)

2.1. Flow Control

In microfluidics, the sheath fluids and sample fluid can be considered as Newtonian fluids, which are continuous, laminar and incompressible. In microflow cytometers, passive flow driven by capillary force or gravity or active pumps driven by an external power source are used to provide continuous flow through the devices [21,22]. Since the cross-section of the channel on the scale of a few 10 s of micrometers and the sidewalls are smooth, the flow in the microchannels can be classified as a Stokes flow with a Reynolds number less than one, meaning that the flow in the channel is in the laminar regime. Small dimensions of the microchannel may raise the risk of clogging by big particles or cells, clumps of cells or even extraneous debris. Details about basic concepts, fabrication strategies and advanced applications of hydrodynamic focusing in microflow cytometers can be found in a review by Ainla *et al.* [23].

2.1.1. 2D Hydrodynamic Flow Focusing

The hydrodynamic focusing technique used in both benchtop and microflow cytometers is one of the most successful and ubiquitous flow focusing techniques. The sample fluid is sandwiched between a sheath fluid in both sides, and since the Reynolds number is low and the fluids are in the laminar regime, there will be no turbulent mixing of the fluids. Figure 2a shows a typical structure used to achieve 2D hydrodynamic focusing in a microflow cytometer [24]. Similar structures that narrow the sample fluid between two sheath fluids have been widely used in microflow cytometry [16,25–29]. The width of the focused sample stream is related to the ratio of sample to sheath flow rate and effectively allows the user to tailor the sample stream width to the application's requirements [30]. The sample stream's width must be large enough to accommodate the largest particles in the sample population, yet not too large as to allow particles to flow side-by-side in the sample stream. It must be noted that the vertical channel height defines the height of sample fluid in 2D hydrodynamic scheme, and thus, careful consideration must be taken as to the channel height and the characteristic size of the cells or particles under inspection.

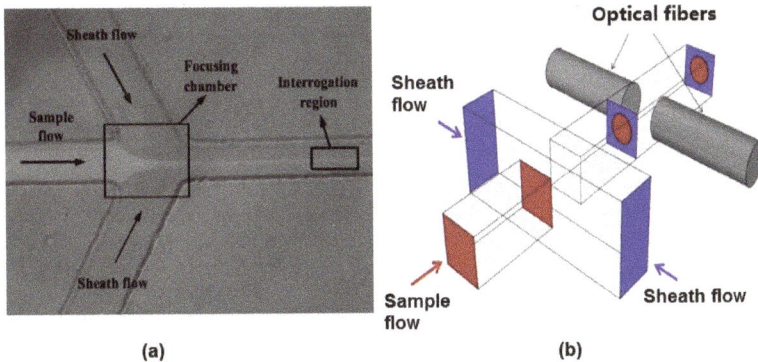

Figure 2. Reported hydrodynamic focusing methods in a microflow cytometer. (**a**) A typical structure of 2D hydrodynamic focusing. Reproduced from [24]. (**b**) A straight forward 3D hydrodynamic focusing structure. Reprinted from [31] with kind permission from OSA Publishing.

2.1.2. 3D Hydrodynamic Flow Focusing

3D hydrodynamic focusing confines the sample fluid to the center of the microchannel in both vertical and horizontal directions, an improvement on the inability of 2D hydrodynamic focusing to focus fluid in the channel's vertical dimension. The straightforward way to achieve 3D hydrodynamic focusing is to use deeper orthogonal sheath fluids, as shown in Figure 2b [31]. Two lateral fluids get in above and below the sample fluid in addition to lateral directions and push the sample fluid from both vertical and horizontal directions in a microchannel with a larger dimension than that of the sample channel. Since deeper channels are difficult to fabricate, requiring three fabrication and two alignment steps, 2D hydrodynamic focusing and its one fabrication step is preferred. However, a strategy of using 2D hydrodynamic focusing twice has been applied to achieve 3D hydrodynamic focusing. By using a planar structure, two sheath fluids A and B can be used to focus the sample fluid vertically, and the sample fluid was focused horizontally by another sheath fluid, C [32]. Experimental results and numerical simulation results show that the sample stream is focused to a small region in the center of the microchannel.

With 3D microfabrication technology, more complex structures are fabricated to focus the particles in two dimensions, such as oblique cylinders or grooves [10]. Sundararajan *et al.* used the "membrane sandwich" method, which contained two sheath fluids from lateral directions and another two on top and at the bottom stacked on the inlet point to create a 3D hydrodynamic focusing microchip [33]. Hairer *et al.* focused the sample fluid by using three sheath fluids in a non-coaxial sheath flow device [34]. V-shaped or chevron-shaped grooves were fabricated in microchannels to focus the sample fluid in both lateral and vertical directions [35,36]. Similar V-shaped slants were also applied in 3D mixing [37]. More recently, Nawaz *et al.* achieved 3D hydrodynamic focusing by using microfluidic drifting with different curvature angles [38].

2.1.3. Other Methods

Besides hydrodynamic focusing, acoustics [39–43], dielectrophoresis (DEP) force [44–47], electrokinetics and magnetophoresis (MAP) [48–50] can be used alternatively to focus particles and cells in a microflow cytometer. Acoustic-based focusing methods do not need sheath fluids for 3D focusing. The standing surface acoustic waves (SSAW) field generated by two parallel interdigital transducers (IDTs) applies lateral and vertical acoustic radiation force to the particles or cells, as shown in Figure 3a. Particles or cells are focused in the center of the microchannels where the pressure node is located. Recently, Chen *et al.* created an SSAW-based 3D focusing microflow cytometer [41], as shown in Figure 3b.

Dielectrophoresis (DEP) is another approach to achieve 3D focusing in microchannels without sheath fluids. In DEP, a non-uniform oscillating electric field creates a dipole on the cells or particles that will experience a negative or positive force depending on the dipole's phase with the applied AC field and the strength of the electric field at each end of the dipole. By changing the frequency of the field, it is possible to tune the strength of the force on the particles or even to switch the direction of the DEP force. The DEP force can adjust particle's or cell's equilibrium position-normally at the center of the channel in the vertical position by utilizing a pair of parallel microelectrodes on the top and bottom surface of the microchannel. Usually, particles or cells experience a negative force as a positive DEP force pulls the particles or cells towards the surface of the electrode where the greatest field gradient occurs that could destroy the cells [51]. The DEP force depends on the size and electrical properties of the particles or cells, the electrical properties of the sample fluids and the electric field. Large particles or cell move slower than smaller particles or cells, and the focusing pattern of each cell or particle is different. Many microchip-based flow cytometers now use more than one technique to confine sample flow. Lin *et al.* [52] combined the 2D hydrodynamic focusing and DEP method to obtain 3D focusing: two electrodes exerted a DEP force from the vertical direction on the particles and cells, which had already been focused laterally hydrodynamically by two sheath fluids. More recently, Zhang *et al.* presented a novel DEP-inertial microflow cytometer, which combined the DEP force and inertial force to achieve vertical-focusing [46]. The MAP theory is similar to DEP force, except that the electric field is replaced by the magnetic field. In addition, particles or cells need to be attached to a magnetic bead so that they can move to the interrogation point exactly.

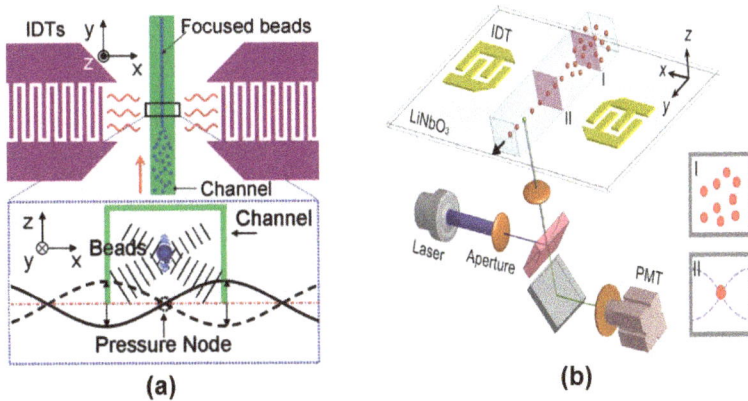

Figure 3. (**a**) A schematic diagram of standing surface acoustic waves (SSAW) focusing. The pressure node located at the center of the microchannel was generated by the SSAW field created by two parallel interdigital transducers (IDTs). When particles or cells enter the SSAW field, the acoustic radiation force (vertically and horizontally) moves the particles to the pressure node. Reprinted from [39] with kind permission from Royal Society of Chemistry. (**b**) A schematic diagram of a SSAW-based microflow cytometer. Reprinted from [41] with kind permission from Royal Society of Chemistry.

2.2. Light Guide and Collection

Particles or cells focused in the center of the microfluidic channel are interrogated by a light beam. In a conventional flow cytometer, a light beam is guided to the capillary tube for excitation, while the various light signals are collected by an objective lens and subsequently split and detected in free-space by bulky optical lenses, dichroic mirrors and other components. A first step towards an optofluidic microflow cytometer involved moving the excitation optics to the chip. Integrating lenses onto the chip eliminates the need for free-space optical alignment while reducing the size of the microflow cytometer device and making the device more portable and durable. Free-space light collection was very similar to the conventional flow cytometer, as shown in Figure 1 [20]. This section will focus on the simulation and design of integrated on-chip optical systems for optical excitation in an optofluidic microchip-based flow cytometer.

2.2.1. Excitation Sources and Optical Fibers

As shown in Figure 1, a laser beam is coupled into a fiber, and the fiber subsequently couples light to an integrated waveguides on the chip to precisely deliver the light to the channel to excite the FL of the particles or cells and generate the scatter signals [20]. In some research papers, light-emitting diodes (LEDs) and laser diodes are used as a source, and either could be used; however, light from an LED is noncoherent light with a wide bandwidth, but has a low cost, while the light from a laser diode is coherent light with a narrow bandwidth and has a higher cost. The correct source can be selected based on the application.

Optical fibers (both single-mode fibers and multi-mode fibers) can be coupled with lasers to provide decent beam shaping at the interrogation region; however, the beam diverges as it leaves the guiding medium. Optofluidic devices coupled with single-mode fibers have been well-reviewed by Blue *et al.* [53]. A device with an integrated on-chip lens system couples light from an optical fiber to an on-chip waveguide to deliver the light to the lens system and focuses and shapes the light in the interrogation region [54,55]. Instead of integrated waveguides on the chip, microgrooves fabricated in the functional layer can help embed optical fibers into the microchip eliminating the need to align fibers to the chip. These inserted fibers can be used for the collection of light signals, as well. Inserted fibers ensure a complete optically-guided approach, from source to detector [2,26,56].

Matteucci *et al.* [56] fabricated grooves for the insertion of optical fibers to achieve precise alignment of optical power (as shown in Figure 4a).

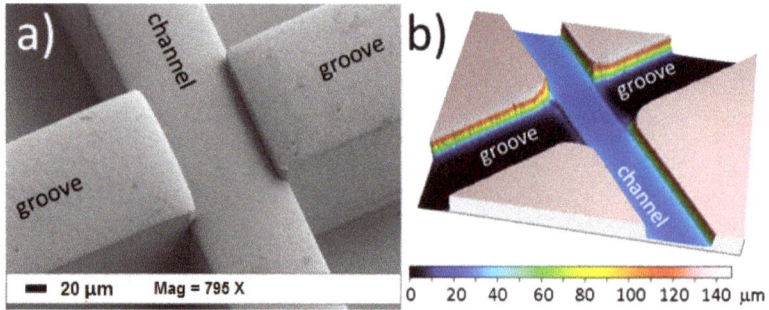

Figure 4. (**a**) SEM microgram of microgrooves for optical fibers. (**b**) confocal microscope profilometry of the microchip that shows the depths of microchannel and microgrooves. Reproduced from [56] from *Micromachines* published by MDPI.

2.2.2. Waveguides

Optofluidic waveguides, such as solid-core/liquid-cladding waveguides (SCLC), liquid-core waveguides (LCW) [57,58] and hybrid core waveguides (HCW) [59,60], have been reported. Guiding of the light is ensured if the refractive index of the core material is higher than that of the cladding. In an optofluidic microflow cytometer, deionized (DI) water, water-based liquids and organic-based liquids are typically used for cladding materials, while glasses, polymers and semiconductors are typical materials used for the core. Choi *et al.* [61] used DI water as a core fluid and 2,2,2-trifluoroethanol as cladding fluid to form a waveguide. Liquid-core/air-cladding (LA) waveguides have been integrated into an optofluidic device by Lim *et al.* [62]. Shi *et al.* [60] demonstrated a hybrid waveguide consisting of a liquid-liquid waveguide and a liquid-solid waveguide to achieve real-time self-imaging in a microchannel. Compared to 2D liquid-liquid waveguide, the 3D liquid-liquid waveguide is surrounded by cladding fluid in both directions, and the confinement of light is better [63]. Yang *et al.* demonstrated bending and manipulating light via optofluidic waveguides with their unique optical properties [64].

Optically transparent photoresists, such as SU-8, are widely used to fabricate on-chip waveguides integrated simultaneously with the microfluidic channel during the fabrication process [54]. SU-8 can function as the core while voids provide air to function as cladding material. As the refractive index of SU-8 (about 1.59) is higher than that of the air, strong optical confinement is observed, lowering the background noise [65]. Figure 5a shows a typical air cladding SU-8 core waveguide fabricated by Watts *et al.* [65]. Light is guided through the long straight waveguide to the lens system, shaping in the microchannel and then collected by multiple waveguides at angles at 5°, 30° and 75° to the input laser beam axis. FSC, FL and SSC are collected from each of the angled waveguides (respectively), where the angled on-chip waveguides help to avoid noise due to a couple of stray light signals from the input laser diode. To further improve the signal to noise ratio (SNR), an angled input waveguide along with an angled lens system were used to reduce the noise by allowing a full 90° angle between the input and SSC, as shown in Figure 5b. A low background noise is created for SSC, which has the same wavelength as that of the excitation beam. Waveguides made by optically transparent polymers, such as poly(dimethylsiloxane) (PDMS), can also be integrated onto the device. In addition to polymers and photoresists, researchers have shown that other materials can also function as waveguides. For example, Emile *et al.* [66] used 1D soap films as waveguides to guide light coming from a laser diode.

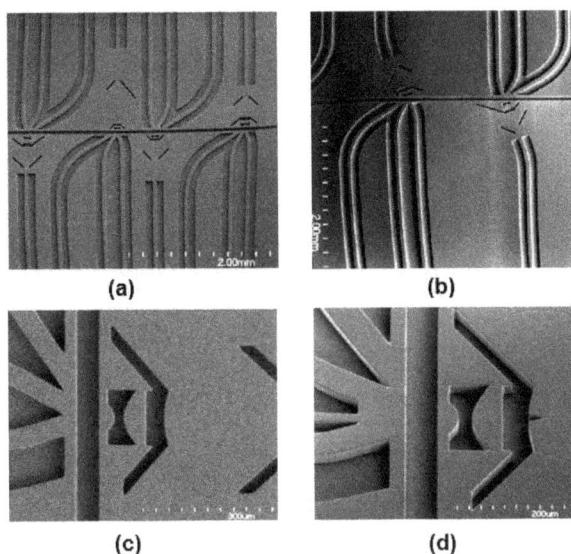

Figure 5. (a) SEM images of four microflow cytometers with different optical systems. Long straight waveguides direct lights to the lens system; waveguides at different angles are used to collect signals. (b) Images of angled input waveguides and the lens system to reduce background noise for SSC and FSC detection. (c) SEM images of on-chip air lens system without notches. (d) SEM images of on-chip air lens system with notches. Reproduced from Watts [65] with permission.

2.2.3. On-Chip Lens System

Due to the large numerical aperture of the on-chip waveguide inherit in the materials' large index contrast, the beam coming out of the waveguide will diverge and expand as it traverses the distance from the waveguide facet to the interrogation region. In other words, the light that propagates to the interrogation region will have a large spot size and poor uniformity. For an optofluidic microflow cytometer, on-chip lens systems can be integrated in the microchip to replace bulky and expensive optical components used in free-space solutions. Beam shaping is used to focus the laser beam via a 2D lens system embedded on the microchip between the waveguide and the microchannel, increasing the uniformity of the light for interrogation (Figure 5). Watts *et al.* [54,55,65,67–70] reshaped the beam from the excitation laser to an optimized geometry in the interrogation region to enhance detection. Beam shaping process specifically reshaped the input laser spot geometry to a designed spot geometry where the center portion of the laser beam was altered to obtain a much smaller and uniform beam. For example, by adjusting design parameters of each surface in the lens system, a beam size of 1.5 μm and 3.6 μm, defined by the full width at half-maximum (FWHM), was formed at the focusing point in the microchannel [65]. The initial beam waist was around 50 μm, which was almost 33 times larger than the reshaped beam waist, and the added benefit of a significant corresponding increase in the beam intensity is also achieved. Simulations using commercial ZEMAX software (2005) shown in Figure 6a demonstrate the process of the shaping the beam from the input to output. Light emitted from the laser passes through the SU-8 waveguide, propagates along the SU-8/air lenses, and forms beam waists of 3.6 μm and 10 μm in the two examples shown. Fluorescent images of the beam with and without shaping (Figure 6b,c, respectively) show that the waveguide without integrated lenses have no control over the beam geometry. It is worth noting that the measured coefficient of variation (CV) of fluorescent beads was strongly dependent on the beam geometry and bead sizes: 2.5-μm fluorescent beads had the best CV of 8.5% for a 3.6-μm beam waist [65].

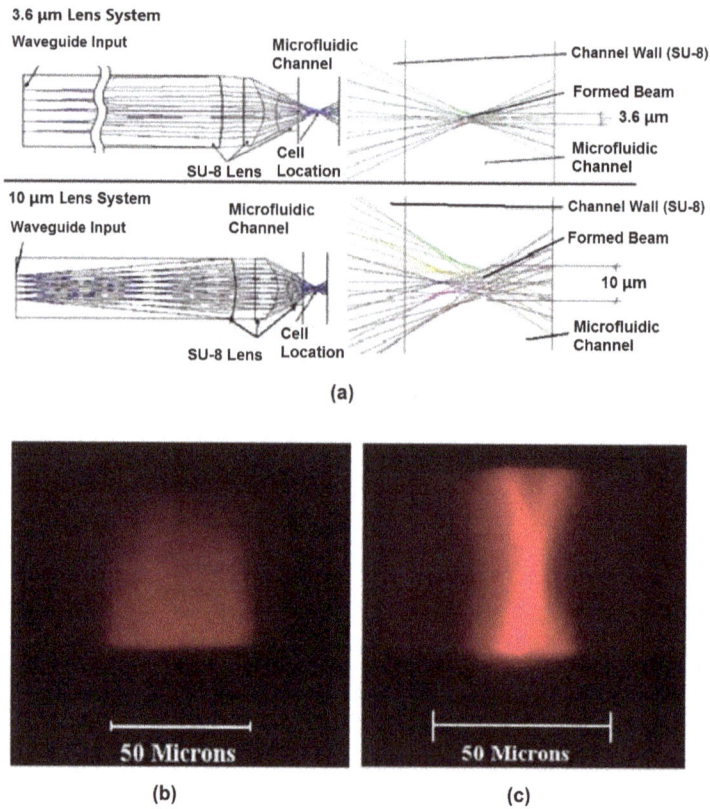

Figure 6. Beam shaping simulation results and fluorescent images of formed beam shape. (**a**) ZEMAX simulation results of 3.6-μm and 10-μm on-chip beam shaping lens systems. (**b**) Fluorescent image of the input excitation beam input directly from a waveguide without any lens. (**c**) Fluorescent image of formed beam shape after passing through a 10-μm beam shaping lens system. Reprinted from [67] with permission from OSA Publishing.

In conventional flow cytometry, a thin obscuration bar located before the detector, onto which the input beam is focused, is used to block the laser beam from reaching the detector directly. This technique is done because the laser beam propagates along the same axis that the FSC and obscures the FSC signal. In an optofluidic device, a notch was applied in the first surface of the lens system, functioning as an obscuration bar, as seen in Figure 7 [65]. The notched lens system forms a dark spot on the facet of a collection waveguide without influence on the beam geometry in the interrogation region. This notched design enhanced the SNR and improved the reliability of on-chip detection for FSC: results show that a false positive rate as low as 0.4% can be achieved [70].

In some optofluidic microflow cytometers, researchers combined the advantages of both liquid and light. Tang *et al.* [18] created a reconfigurable liquid-core/liquid-cladding lens (L^2 lens) formed by three laminar flows. Very similar to the air lens system, two streams of a lower refractive index function as the cladding, and a stream of a higher refractive index functions as the core. The focal length of the lens can be changed in real time by changing the relative flow rates of the three streams without mechanical moving parts. At the same time, the liquid lenses provide an optically-smooth interface for light manipulation. Those novel optofluidic components provide new opportunities for on-chip flow cytometers and cross the boundary of multiple disciplines.

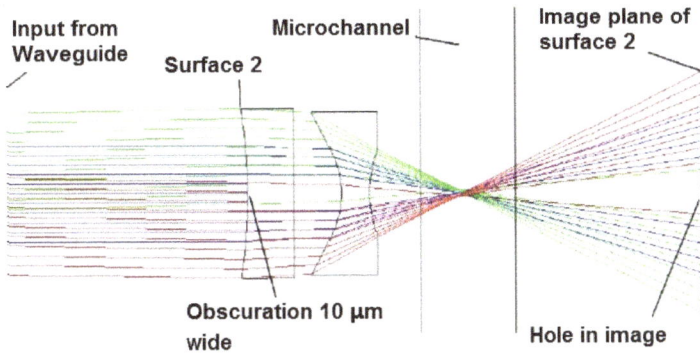

Figure 7. ZEMAX simulations for a 3-μm lens system that inserts a notch for forward scattered light detection. Note how the notch is re-imaged on the waveguide behind the channel. Reprinted from [65] with permission.

To minimize losses in the system, it is important to match the size of the on-chip waveguide and coupling fiber. The roughness of the waveguide, lens surfaces and channel wall can also cause propagation losses in the device. The SEM image in Figure 8 shows that the sidewall and facets are very smooth. The excellent quality of the waveguide facet, channel wall and lens surfaces allow low amounts of escaping light due to scattering from imperfection in the photolithographically-formed side walls. To further improve the sensitivity of the device, higher input laser power can be provided to enhance the SSC or FL signal of the particles or cells. The effect of roughness of the on-chip lens system can be ignored due to the high signal-to-noise ratio.

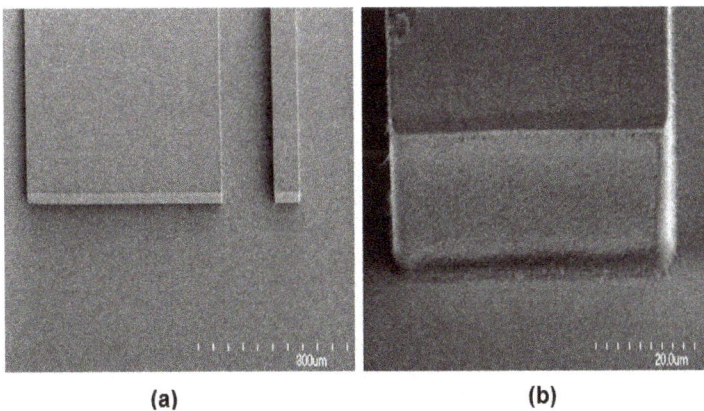

(a)   **(b)**

Figure 8. (a) SEM image of a waveguide facet. (b) A close-up SEM image of a waveguide facet showing the smooth optical coupling face. Reprinted from [65] with permission.

2.3. Data Collection

Scattered light and fluorescent light produced by the interaction between the light beam and particles or cells are detected by the optical detectors connected to a computer. The most common optical detectors are photomultiplier tubes (PMTs), avalanche photodiodes (APDs), P-doped/Intrinsic/N-doped (PIN) photodiodes [21,26], charge-coupled device (CCD) cameras and CMOS arrays [71–73]. PMTs have been widely used in commercial flow cytometers due to the high sensitivity and reliability, especially when dichroic mirrors are used to split the beam for multi-color fluorescence detection. PMTs and other photodiodes can convert incident light to an electrical current and multiply it by as much as 10^8 times. Recently, compact PMTs have been made available on the commercial market, which can be used to replace the old large PMTs and to reduce the overall size of the detection system. Compared to PMTs, APDs are sensitive to temperature, and PIN photodiodes have a much simpler structure. PIN photodiodes can replace PMTs and APDs to reduce the cost when the light signal is strong. PIN photodiodes with lock-in amplification can be utilized for the single cell or particle fluorescence detection [26]. Kettlitz *et al.* avoid expensive PMTs and substitute them with a PIN photodiode, achieving a maximum particle detection frequency of 600 particles/s [74].

Compared to photodiode detectors, CCD camera and CMOS imaging can provide instant images, but the speed of the fluid is limited [51]. Hoera *et al.* utilized a CCD camera for fluorescence imaging of temperature and reaction process in a microfluidic chip reactor [75]. Yang *et al.* conducted an experiment at a flow rate of 10 µL/h to allow CCD cameras to capture the accurate images for real-time cell separation in a microflow cytometer [72]. To further improve the sensitivity of the detector and reduce the cost of the microflow cytometer, Eyer *et al.* [73], added titanium dioxide (TiO$_2$) particles into PDMS to increase the light signal intensity from the interaction between particles and light source, improving the sensitivity of CCD cameras indirectly.

2.4. Data Analysis

As shown in Figure 2, light signals collected by optical detectors are further amplified by a current-to-voltage amplifier, and then, the voltage signals are digitized by a data acquisition card. LabView programs can be utilized to record the data for further analysis [20]. Data analysis is usually performed using customized MATLAB codes [20,76].

When a cell or a particle passes through the interrogation region, a burst of light will generate an electrical pulse in the analysis system. The pulse duration relates to flow rate, the beam waist and the size of cell or particle. FSC intensity is proportional to the size of the cell or particle, and SSC intensity depends on the granularity. Each pulse is characterized by FSC, SSC, single-color or multi-color FL and pulse duration. The average intensity and pulse duration are typically calculated. To remove some background noise or internal PMT noise, a threshold is set. Figure 9a shows the data analysis results of a of mixture beads and cells flowing in an optofluidic microflow cytometer [20]. One dashed threshold is set to distinguish 2- and 4-µm beads; another dotted threshold is set to separate beads and *Escherichia coli* cells.

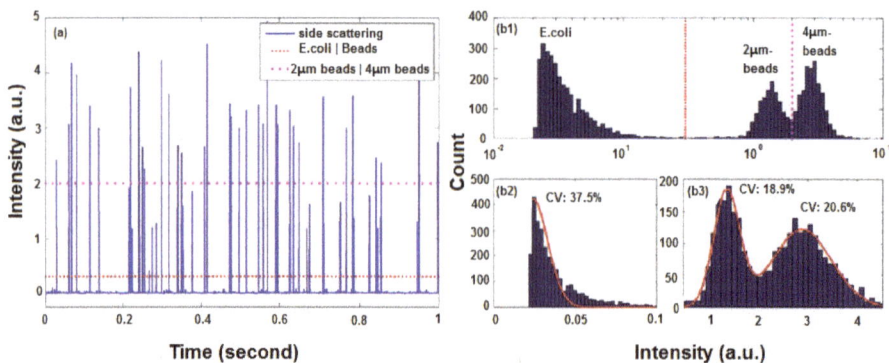

Figure 9. Data analysis results of a mixture of beads and cells flowing in an optofluidic microflow cytometer. (**a**) One second raw data of SSC signal intensity from a test with a mixture of *E. coli* cells and beads of 2 µm and 4 µm in diameter. (**b1**) Statistical histograms of events by total beads and *E. coli* cells with intensity on a logarithmic scale. (**b2**) Statistical histograms of events produced by *E. coli* cells on a linear scale and its Gaussian fitting. (**b3**) Statistical histograms of events by beads with intensity on a linear scale and Gaussian fittings. Reprinted from [20] with permission from Wiley.

In flow cytometry, simultaneous detection of multiple parameters is the source of its analytical power. By analyzing and comparing the scattered light signals and fluorescent light signals of a single-cell or single-particle against the total population, it is possible to see the similarities and differences for further counting and identification. Figure 9b shows histograms of events with intensity on a logarithmic scale and a linear scale. Cells and beads of different diameters show various distribution features, which can be fitted by Gaussian curves with different coefficients of variation (CV). For a microflow cytometer with multi-color fluorescence channels, a multi-parameter plot can be used to do further analysis.

3. Fabrication and Integration

3.1. Materials

Traditionally, silicon and glass substrates are the most common materials used in microflow cytometers. Recently, inorganic materials, like ceramics, and polymers, like PDMS, and even paper [77] have been used to construct microfluidic devices. Silicon and glass technologies, as well as polymer technology have been reviewed in many papers [12,77,78].

Wet or dry etching methods are applied to create microstructures on a silicon or glass substrate, but organic long-chain polymers are attracting more and more attention with the growing interest in fabricating multilayered structures. Polymers are less expensive and convenient for mass production. More importantly, most polymers are optically transparent to visible wavelengths of light and adaptable through chemical modification for bonding to glass or silicon substrates. Polymers can be divided into elastomers and thermoplastics. PDMS, as one of the elastomers, was first used as a substrate in the late 1990s. Since then, PDMS has established itself as the most commonly-used elastomer in microfluidics. Chemical modification of PDMS has allowed the diversification of its application in microfluidics. PDMS structures can be cured on molds at room temperature for microchannels or other microstructures, and PDMS can provide good sealing properties after chemical modification. Zhang *et al.* [79] sealed SU-8 microfluidic channels using PDMS after the N_2 plasma treatment. Amino groups generated by N_2 plasma on the PDMS surface reacted with the residual epoxy groups on the SU-8 surface. The bond was long-term resistant to water, and the structure could withstand a high degree of stress. Polystyrene (PS), polycarbonate (PC), poly(methyl methacrylate) (PMMA) and cyclic olefin copolymer (COC) are other thermoplastic polymers that are used in microfluidics.

As stated earlier, SU-8 is a commonly-used epoxy-based negative photoresist that can be utilized to form the functional layer (waveguides or lenses) when processed on a substrate [24,67,80]. High aspect ratio structures can be obtained in SU-8 by lithography. Researchers now are seeking methods to integrate PDMS and SU-8 together to take advantage of both materials. Ren *et al.* [80] bonded SU-8 and PDMS using the aminosilane-mediated bonding method in a microfluidic device for neuroscience research. Paper is a promising new material with low cost and easy fabrication process. Furthermore, paper is available everywhere and relatively environmentally-friendly. Liu and Crook [77] fabricated a 3D paper microfluidic devices simply by hand folding. Colorimetric and fluorescence detection of glucose and protein were achieved.

3.2. Device Integration

A multilayered PDMS/SU-8 devices is commonly used in an optofluidic microflow cytometer. Standard techniques, like wet and dry etching on glass or silicon substrates, have been replaced by soft lithography, photolithography and different bonding techniques in recent decades. Figure 10 shows the individual layers for the integration of the device, and the detailed process can be seen in Figure 11. The PDMS layer is fabricated by PDMS molding and functioned as a upper layer to seal the fluidics and to form an upper optical cladding layer. The SU-8 layer is patterned on a silicon or glass substrate by exposure to UV light through photomasks. Due to the high aspect ratio, the waveguides and optical components can be integrated on the SU-8 layer. To assemble the device, the PDMS layer is bonded to the SU-8 layer after a plasma treatment [67,80]. The PDMS cover provides a complete seal, and extra glass pads are bonded on the top surface of PDMS. The sandwiched device is rigid enough and can withstand a high pressure. Custom optical and fluidic components can be integrated on a small chip to achieve specialized and efficient function.

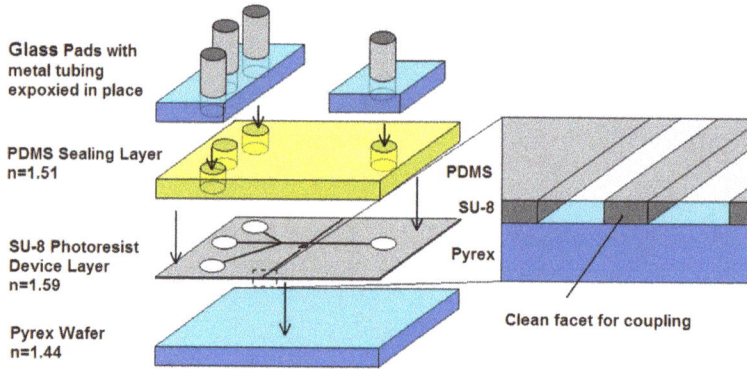

Figure 10. A schematic diagram showing the integration of a multilayered PDMS/SU-8 device. SU-8 is the functional layer; PDMS covers and seals the device; and glass pads allow solid fluidic interconnects. Reproduced from [65] with permission.

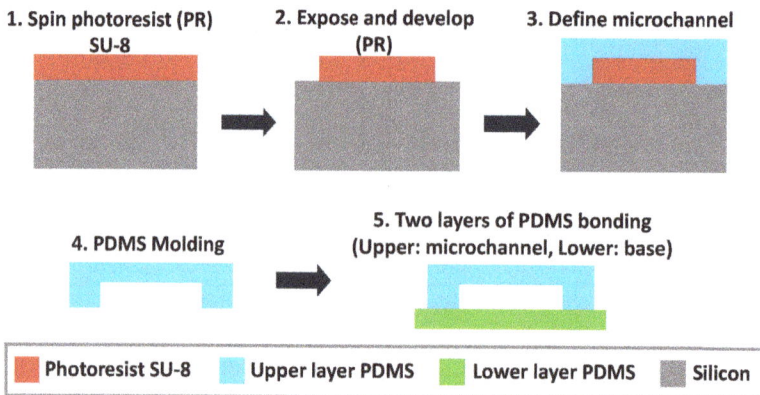

Figure 11. Standard fabrication procedure of a PDMS/SU-8 device. Reprinted from [81] from *Micromachines* published by MDPI.

4. Performance of Optofluidic Microflow Cytometers

Optofluidic microflow cytometers aim to provide a compact and automated method capable of high-throughput screening, low reagent consumption, high sensitivity and high selectivity for small particles or cells detection. Current researchers are moving forward step by step to achieve those goals and narrowing the performance difference between the conventional benchtop and optofluidic microflow cytometers.

The sensitivity of the optofluidic microflow cytometer is improved by the on-chip lens system. The fluorescence sensitivity of flow cytometers depends on the background noise, effective intensity of light signal and the detection efficiency. SNR is the ability to resolve a pulse from the noise. The minimum SNR ratio for a reliable detection is three, while the SNR value of fluorescent beads measurement was 80–300 from devices by using novel on-chip lens systems and notched designs to reduce the background noise [68].

Table 3 shows the parameters of optofluidic microflow cytometers developed recently and their performance measured by the coefficient of variation (CV). The CV is defined as the ratio of the standard deviation to the mean of the light signal intensity expressed as a percentage. This effectively measures the dispersion of intensity of the detection events [82]. A smaller CV indicates that there is less error introduced by the actual device and that identical samples will have identical detected signals, meaning that the microflow cytometer has a higher ability to differentiate the slight differences of the type of particles or cells in the entire population.

The throughput of optofluidic microflow cytometer varies from 30 particles or cells/s to 2000 cells/s or even to 50,000 cells/s. Different levels of throughputs are achievable depending on the specific application. For *E. coli* or bacteria detection, where the concentrations of targeted cells are very low, a low throughput is used. For rare cell analysis or blood cell counting, high throughputs are required. As shown in Table 3, the CV values for cells are significantly higher than those of blank or fluorescent labeled beads. Since the refractive index contrast between cells and sample fluid (usually water or phosphate-buffered saline) is small, the intensity of scattered light is not as strong as that of beads. Fluorescent labeling is applied to provide an easily detectable parameter to help improve SSC detection by correlating the two parameters; if an FL is detected, then a SSC must be detected, as well. This can indirectly enhance the SNR of the SSC. The CV value of SSC intensity produced by *E. coli* cells was 37.5% [20], while the CV value of FL intensity produced by labeled human embryonic kidney (HEK) cells is 13.4% [38].

An optofluidic microflow cytometer with typical 2D hydrodynamic focusing, on-chip beam shaping and collection was studied by Watts *et al.* [55]. Scattered light signals of 1-, 2- and 5-μm blank beads were collected by on-chip waveguide, obtaining CV values of 16.4%, 11.0% and 12.5%, respectively. It also showed that the performance of optofluidic microflow cytometers is based on the combination of the beam geometry used and bead size. Barat *et al.* [2] demonstrated a 2D hydrodynamic focusing optofluidic microflow cytometer with an on-chip lens system for light guiding and grooves for inserted optical fibers for collection. The flow cytometer successfully differentiates 10–25-μm beads based on fluorescence and scattered light. The best CV value was 4.9% for the side scattered light intensity of 25-μm beads.

The performance of a 2D hydrodynamic focusing optofluidic microflow cytometer with free-space collection was determined to be very comparable to conventional cytometers. Watts *et al.* [54] focused the beam waist to 6 μm, and fluorescent signals from 2.5-μm beads showed a superior CV of 9.03%. Mu *et al.* [24] detected labeled *E. coli* cells, achieving a detection efficiency of 89.7% and 94.5% for fluorescence signals and scattered light signals, respectively. The detection accuracy was 84.3% and 88.8% for fluorescence and scattered light detection, respectively, as compared to the standard haemocytometer method.

Nawaz *et al.* [38] presented a novel microfluidic drifting-based 3D hydrodynamic focusing optofluidic microflow cytometer with free space light collection. The best CV of 2.37% for fluorescent beads was achieved, which is comparable to a commercial benchtop flow cytometer. Frankowski *et al.* [28,29] developed two microfluidic sensors based on optical and impedance analysis both operating simultaneously. They demonstrated a superior CV of 3.2% for 8.12-μm beads. A combination of multiple methods or techniques is a good way to improve the sensitivity of microfluidic devices.

Table 3. Performance of recently developed optofluidic microflow cytometers.

Flow Control	Beam Shaping	Light Collection	Sample	CV of SSC (%)	CV of FL (%)	Throughput (Cells or Particles/s)	Ref.
2D HF	Yes	Free-space	E. coli	37.5	–	~101	[20]
2D HF	Yes	On-chip	2 µm beads	11	–	~30	[55]
2D HF	Yes	On-chip	15 µm beads	12	17.1	~100	[2]
2D HF	No	Free-space	Labeled E. coli	36.2	30.7	~350	[24]
2D HF	No	Free-space	1 µm beads	14.95	24.73	~83	[24]
2D HF	Yes	Free-space	2.5 µm beads	–	9.0	~28	[54]
3D HF	No	On-chip	10 µm beads	12	8.3	-	[31]
3D HF (cascade focusing)	No	Free-space	8.12 µm beads	–	3.2	–	[29]
3D HF (microfluidic drifting)	No	Free-space	1.9 µm beads	–	2.4	~2163	[38]
3D HF (microfluidic drifting)	No	Free-space	HEK 293 cells	–	13.4	–	[38]
3D HF (cascade focusing)	No	Free-space	Beads	–	3.0	–	[28]
3D SSAW	No	Free-space	HL–60 cells	–	22.0	–	[41]
3D SSAW	No	Free-space	7 µm beads	–	19.4	~772	[41]
3D SSAW	No	Free-space	10 µm beads	–	10.9	~537	[41]

Typically, good 3D hydrodynamic focusing ability in both lateral and vertical directions can achieve a lower CV than 2D focusing. However, beam shaping can also improve the detection efficiency and obtain a lower CV value even with 2D hydrodynamic focusing. A CV of 15.9% for 2-µm beads was achieved by Watts *et al.* [55] using 2D hydrodynamic focusing, while a similar CV of 15.4% for 10-µm beads was obtained in a 3D hydrodynamic focusing made by Testa *et al.* [31]. This is because Watts *et al.* used an on-chip lens system to focus the beam waist to 1.5 µm and formed a superior uniform region of light intensity of the interrogation region [55]. Both devices provide free optical alignment, but the grooved structure in Testa's device is easier to fabricate.

Besides hydrodynamic focusing techniques, SSAW also provides a good focusing quality. A mixture of 7-µm beads and 10-µm beads was distinguished by an SSAW-based microfluidic cytometer studied by Chen *et al.* [41], with CVs of 19.4% and 10.9%, respectively. Fluorescently-labeled human promyelocytic leukemia cells (HL-60) were successfully detected with a CV of 22.0%.

The typical CV value achievable by the conventional benchtop flow cytometer is about 5%–15%. The performance of a optofluidic microflow cytometer has developed dramatically from 24.73% by Mu *et al.* [24] to 3% by Frankowski *et al.* [29] in the past few years. The performance of an optofluidic microflow cytometer now can be comparable to the conventional benchtop flow cytometer. The improvement of the device's performance is attributable to the development of microchip devices' enhanced flow control methods, the on-chip lens system that provides custom and robust beam shaping capabilities, the signal collection method and the rapidly-growing microfabrication techniques.

5. Conclusions and Future Perspective

Optofluidic microflow cytometers are attracting more and more research attention, combining multiple fields and disciplines in a microchip. Optofluidic microflow cytometers integrated with optics provide significant enhancements for flow cytometry. Low costs, higher sensitivity, free optical alignment and smooth interaction interfaces are the obvious advantages. Optofluidic microflow cytometers offer significantly lower costs and size reductions, as well as low reagent requirements and portability advantages over a benchtop flow cytometer.

Various attempts have been made to integrate on-chip waveguides or grooves for the guided insertion of optical fibers into the microchip. Besides on-chip lenses or waveguides fabricated by SU-8 or PDMS, novel types of liquid-core/liquid-cladding lenses, liquid-core/liquid-cladding waveguides and hybrid core waveguides have been used in optofluidic devices. The integrated lens system helps shape the beam emitted from the excitation source, providing an optimal geometry and uniform region for interaction with the specimen. By taking advantage of both optics and fluidics, the performance of optofluidic microflow cytometers has advanced to a point where devices are comparable to that of a conventional flow cytometer in the past few years. To the extent of our knowledge, the best CV value achieved for a microflow cytometer was less than 3% [29,38].

Currently, the optofluidic devices have successfully miniaturized the optical and fluidic components of the flow cytometer, while the miniaturization of the whole system is still challenging. System integration needs to have more attention paid to it. External syringe pumps, light collection detectors and other components still somewhat limit the portability of the optofluidic microflow cytometer. The miniaturization of these components needs to join the current stat of the optical and microfluidic control components for continuous development of future POC applications. For POC diagnostic and other field applications, a simple method for mass production should be invented in the future to further reduce the cost. For commercialization, optofluidic microflow cytometers need to provide significant operational advantage over conventional flow cytometers. A fully-portable, low cost, easy to operate and effective optofluidic microflow cytometer can be reasonably expected in both the research field and in the commercial market.

Acknowledgments: This research is partially supported by an Ontaio Research Fund (ORF) grant (Grant No. RE-WR-10) and an NSERC Discovery grant (Grant No. RGPIN/262023-2013).

Author Contributions: Changqing Xu and Zhiyi Zhang conceived of and designed the structure of this review article. Yushan Zhang collected the references and wrote this paper. Benjamin R. Watts, Tianyi Guo, and Qiyin Fang revised the manuscript.

Conflicts of Interest: The authors declare no conflict of interest.

References

1.   Piyasena, M.E.; Graves, S.W. The intersection of flow cytometry with microfluidics and microfabrication. *Lab Chip* **2014**, *14*, 1044-1059.
2.   Barat, D.; Spencer, D.; Benazzi, G.; Mowlem, M.C.; Morgan, H. Simultaneous high speed optical and impedance analysis of single particles with a microfluidic cytometer. *Lab Chip* **2012**, *12*, 118–126.
3.   Volpatti, L.R.; Yetisen, A.K. Commercialization of microfluidic devices. *Trends Biotechnol.* **2014**, *32*, 347–350.
4.   Psaltis, D.; Quake, S.R.; Yang, C. Developing optofluidic technology through the fusion of microfluidics and optics. *Nature* **2006**, *442*, 381–386.
5.   Horowitz, V.R.; Awschalow, D.D.; Pennathur, S. Optofluidics: Field or technique? *Lab Chip* **2008**, *8*, 1856-1863.
6.   Huang, N.T.; Zhang, H.L.; Chung, M.T.; Seo, J.H.; Kurabayashi, K. Recent advancements in optofluidics-based single-cell analysis: Optical on-chip cellular manipulation, treatment, and property detection. *Lab Chip* **2014**, *14*, 1230–1245.
7.   Crosland-Taylor, P.J. A device for counting small particles suspended in a fluid through a tube. *Nature* **1953**, *171*, 37–38.
8.   Justin, G.A.; Denisin, A.K.; Nasir, M.; Shriver-Lake, L.C.; Golden, J.P.; Ligler, F.S. Hydrodynamic focusing for impedance-based detection of specifically bound microparticles and cells: Implications of fluid dynamics on tunable sensitivity. *Sens. Actuators B Chem.* **2012**, *166-167*, 386–393.
9.   Song, H.; Wang, Y.; Rosano, J.M.; Prabhakarpandian, B.; Garson, C.; Pant, K.; Lai, E. A microfluidic impedance flow cytometer for identification of differentiation state of stem cells. *Lab Chip* **2013**, *13*, 2300–2310.
10. Han, M.; Lee, W.; Lee, S.K.; Lee, S.S. 3D microfabrication with inclined/rotated UV lithography. *Sens. Actuators A Phys.* **2004**, *111*, 14–20.
11. Lee, C.Y.; Chang, C.L.; Wang, Y.N.; Fu, L.M. Microfluidic Mixing: A Review. *Int. J. Mol. Sci.* **2011**, *12*, 3263–3287.
12. Abgrall, P.; Gué, A.M. Lab-on-chip technologies: Making a microfluidic network and coupling it into a complete microsystem–A review. *J. Micromech. Microeng.* **2007**, *17*, R15–R49.
13. Whitesides, G.M. The origins and the future of microfluidics. *Nature* **2006**, *442*, 368–373.
14. Brennan, D.; Justice, J.; Corbett, B.; McCarthy, T.; Galvin, P. Emerging optofluidic technologies for point-of-care genetic analysis systems: A review. *Anal. Bioanal. Chem.* **2009**, *395*, 621–636.
15. Titmarsh, D.M.; Chen, H.; Glass, N.R.; Cooper-White, J.J. Concise review: Microfluidic technology platforms: Poised to accelerate development and translation of stem cell-derived therapies. *Stem Cells Transl. Med.* **2014**, *3*, 81–90.
16. Hashemi, N.; Erickson, J.S.; Golden, J.P.; Jackson, K.M.; Ligler, F.S. Microflow cytometer for optical analysis of phytoplankton. *Biosens. Bioelectron.* **2011**, *26*, 4263–4269.
17. Lei, K.F. Microfluidic Systems for Diagnostic Applications: A Review. *J. Lab. Autom.* **2012**, *17*, 330–347.
18. Tang, S.K.Y.; Stan, C.A.; Whitesides, G.M. Dynamically reconfigurable liquid-core liquid-cladding lens in a microfluidic channel. *Lab Chip* **2008**, *8*, 395–401.
19. Liang, L.; Zuo, Y.F.; Wu, W.; Zhu, X.Q.; Yang, Y. Optofluidic restricted imaging, spectroscopy and counting of nanoparticles by evanescent wave using immiscible liquids. *Lab Chip* **2016**, doi:10.1039/C6LC00078A.
20. Guo, T.; Wei, Y.; Xu, C.; Watts, B.R.; Zhang, Z.; Fang, Q.; Zhang, H.; Selvaganapathy, P.R.; Deen, M.J. Counting of *Escherichia coli* by a microflow cytometer based on a photonic-microfluidic integrated device. *Electrophoresis* **2015**, *36*, 298–304.
21. Huh, D.; Gu, W.; Kamotani, Y.; Grotberg, J.B.; Takayama, S. Microfluidics for flow cytometric analysis of cells and particles. *Physiol. Meas* **2005**, *26*, R73–R98.
22. Lau, A.T.H.; Yip, H.M.; Ng, K.C.C.; Cui, X.; Lam, R.H.W. Dynamics of microvalve operations in integrated microfluidics. *Micromachines* **2014**, *5*, 50–65.
23. Ainla, A.; Jeffries, G.; Jesorka, A. Hydrodynamic flow confinement technology in microfluidic perfusion devices. *Micromachines* **2012**, *3*, 442–461.

24. Mu, C.; Zhang, F.; Zhang, Z.; Lin, M.; Cao, X. Highly efficient dual-channel cytometric-detection of micron-sized particles in microfluidic device. *Sens. Actuators B Chem.* **2011**, *151*, 402–409.

25. McClain, M.A.; Culbertson, C.T.; Jacobson, S.C.; Ramsey, J.M. Flow cytometry of *Escherichia coli* on microfluidic devices. *Anal. Chem.* **2001**, *73*, 5334–5338.

26. Tung, Y.C.; Zhang, M.; Lin, C.T.; Kurabayashi, K.; Skerlos, S.J. PDMS-based opto-fluidic micro flow cytometer with two-color, multi-angle fluorescence detection capability using PIN photodiodes. *Sens. Actuators B Chem.* **2004**, *98*, 356–367.

27. Fu, L.M.; Tsai, C.H.; Lin, C.H. A high-discernment microflow cytometer with microweir structure. *Electrophoresis* **2008**, *29*, 1874–1880.

28. Frankowski, M.; Theisen, J.; Kummrow, A.; Simon, P.; Ragusch, H.; Bock, N.; Schmidt, M.; Neukammer, J. Microflow cytometers with integrated hydrodynamic focusing. *Sensors* **2013**, *13*, 4674–4693.

29. Frankowski, M.; Simon, P.; Bock, N.; El-Hasni, A.; Schnakenberg, U.; Neukammer, J. Simultaneous optical and impedance analysis of single cells: A comparison of two microfluidic sensors with sheath flow focusing. *Eng. Life Sci.* **2015**, *15*, 286–296.

30. Lee, G.B.; Hung, C.I.; Ke, B.J.; Huang, G.R.; Hwei, B.H.; Lai, H.F. Hydrodynamic Focusing for a Micromachined Flow Cytometer. *J. Fluids Eng.* **2001**, *123*, 672.

31. Testa, G.; Persichetti, G.; Bernini, R. Micro flow cytometer with self-aligned 3D hydrodynamic focusing. *Biomed. Opt. Express* **2014**, *6*, 54.

32. Kim, D.S.; Kim, D.S.D.; Han, K.; Yang, W. An efficient 3-dimensional hydrodynamic focusing microfluidic device by means of locally increased aspect ratio. *Microelectron. Eng.* **2009**, *86*, 1343–1346.

33. Sundararajan, N.; Pio, M.S.; Lee, L.P.; Berlin, A.A. Three-dimensional hydrodynamic focusing in polydimethylsiloxane (PDMS) microchannels. *J. Microelectromech. Syst.* **2004**, *13*, 559–567.

34. Hairer, G.; Pärr, G.S.; Svasek, P.; Jachimowicz, A.; Vellekoop, M.J. Investigations of micrometer sample stream profiles in a three-dimensional hydrodynamic focusing device. *Sens. Actuators B Chem.* **2008**, *132*, 518–524.

35. Golden, J.P.; Kim, J.S.; Erickson, J.S.; Hilliard, L.R.; Howell, P.B.; Anderson, G.P.; Nasir, M.; Ligler, F.S. Multi-wavelength microflow cytometer using groove-generated sheath flow. *Lab Chip* **2009**, *9*, 1942–1950.

36. Kim, J.S.; Anderson, G.P.; Erickson, J.S.; Golden, J.P.; Nasir, M.; Ligler, F.S. Multiplexed detection of bacteria and toxins using a microflow cytometer. *Anal. Chem.* **2009**, *81*, 5426–5432.

37. Sato, H.; Yagyu, D.; Ito, S.; Shoji, S. Improved inclined multi-lithography using water as exposure medium and its 3D mixing microchannel application. *Sens. Actuators A Phys.* **2006**, *128*, 183–190.

38. Nawaz, A.A.; Zhang, X.; Mao, X.; Rufo, J.; Lin, S.C.S.; Guo, F.; Zhao, Y.; Lapsley, M.; Li, P.; McCoy, J.P.; *et al.* Sub-micrometer-precision, three-dimensional (3D) hydrodynamic focusing via "microfluidic drifting". *Lab Chip* **2014**, *14*, 415–423.

39. Shi, J.; Mao, X.; Ahmed, D.; Colletti, A.; Huang, T.J. Focusing microparticles in a microfluidic channel with standing surface acoustic waves (SSAW). *Lab Chip* **2007**, *8*, 221–223.

40. Shi, J.; Yazdi, S.; Lin, S.C.S.; Ding, X.; Chiang, I.K.; Sharp, K.; Huang, T.J. Three-dimensional continuous particle focusing in a microfluidic channel via standing surface acoustic waves (SSAW). *Lab Chip* **2011**, *11*, 2319–2324.

41. Chen, Y.; Nawaz, A.A.; Zhao, Y.; Huang, P.H.; McCoy, J.P.; Levine, S.J.; Wang, L.; Huang, T.J. Standing surface acoustic wave (SSAW)-based microfluidic cytometer. *Lab Chip* **2014**, *14*, 916–923.

42. Schmid, L.; Weitz, D.A.; Franke, T. Sorting drops and cells with acoustics: acoustic microfluidic fluorescence-activated cell sorter. *Lab Chip* **2014**, *14*, 3710–3718.

43. Yu, C.; Qian, X.; Chen, Y.; Yu, Q.; Ni, K.; Wang, X. Three-dimensional electro-sonic flow focusing ionization microfluidic chip for mass spectrometry. *Micromachines* **2015**, *6*, 1890–1902.

44. Zhu, J.; Xuan, X. Dielectrophoretic focusing of particles in a microchannel constriction using DC-biased AC flectric fields. *Electrophoresis* **2009**, *30*, 2668–2675.

45. Chu, H.; Doh, I.; Cho, Y.H. A three-dimensional particle focusing channel using the positive dielectrophoresis (pDEP) guided by a dielectric structure between two planar electrodes. *Trans. Korean Soc. Mech. Eng. A* **2009**, *33*, 261–264.

46. Zhang, J.; Yan, S.; Alici, G.; Nguyen, N.T.; Di Carlo, D.; Li, W. Real-time control of inertial focusing in microfluidics using dielectrophoresis (DEP). *RSC Adv.* **2014**, *4*, 62076–62085.

47. Bender, B.F.; Garrell, R.L. Digital microfluidic system with vertical functionality. *Micromachines* **2015**, *6*, 1655–1674.

48. James, C.D.; McClain, J.; Pohl, K.R.; Reuel, N.; Achyuthan, K.E.; Bourdon, C.J.; Rahimian, K.; Galambos, P.C.; Ludwig, G.; Derzon, M.S. High-efficiency magnetic particle focusing using dielectrophoresis and magnetophoresis in a microfluidic device. *J. Micromech. Microeng.* **2010**, *20*, 045015.

49. Zeng, J.; Chen, C.; Vedantam, P.; Brown, V.; Tzeng, T.R.J.; Xuan, X. Three-dimensional magnetic focusing of particles and cells in ferrofluid flow through a straight microchannel. *J. Micromech. Microeng.* **2012**, *22*, 105018.

50. Fernandes, A.C.; Duarte, C.M.; Cardoso, F.A.; Bexiga, R.; Cardoso, S.; Freitas, P.P. Lab-on-chip cytometry based on magnetoresistive sensors for bacteria detection in milk. *Sensors* **2014**, *14*, 15496–15524.

51. Ateya, D.A.; Erickson, J.S.; Howell, P.B.; Hilliard, L.R.; Golden, J.P.; Ligler, F.S. The good, the bad, and the tiny: A review of microflow cytometry. *Anal. Bioanal. Chem.* **2008**, *391*, 1485–1498.

52. Lin, C.H.; Lee, G.B.; Fu, L.M.; Hwey, B.H. Vertical focusing device utilizing dielectrophoretic force and its application on microflow cytometer. *J. Microelectromech. Syst.* **2004**, *13*, 923 – 932.

53. Blue, R.; Dudus, A.; Uttamchandani, D. A Review of Single-Mode Fiber Optofluidics. *IEEE J. Sel. Top. Quantum Electron* **2015**, *22*, 1–12.

54. Watts, B.R.; Kowpak, T.; Zhang, Z.; Xu, C.Q.; Zhu, S.; Cao, X.; Lin, M. Fabrication and performance of a photonic-microfluidic integrated device. *Micromachines* **2012**, *3*, 62–77.

55. Watts, B.R.; Zhang, Z.; Xu, C.Q.; Cao, X.; Lin, M. Scattering detection using a photonic-microfluidic integrated device with on-chip collection capabilities. *Electrophoresis* **2013**, *35*, 271–281.

56. Matteucci, M.; Triches, M.; Nava, G.; Kristensen, A.; Pollard, M.R.; Berg-Sørensen, K.; Taboryski, R.J. Fiber-based, injection-molded optofluidic systems: Improvements in assembly and applications. *Micromachines* **2015**, *6*, 1971–1983.

57. Yalizay, B.; Morova, Y.; Dincer, K.; Ozbakir, Y.; Jonas, A.; Erkey, C.; Kiraz, A.; Akturk, S. Versatile liquid-core optofluidic waveguides fabricated in hydrophobic silica aerogels by femtosecond-laser ablation. *Opt. Mater.* **2015**, *47*, 478–483.

58. Fan, S.K.; Lee, H.P.; Chien, C.C.; Lu, Y.W.; Chiu, Y.; Lin, F.Y. Reconfigurable liquid-core/liquid-cladding optical waveguides with dielectrophoresis-driven virtual microchannels on an electromicrofluidic platform. *Lab Chip* **2016**, *6*, 847–854.

59. Chen, X.; Sakurazawa, A.; Sato, K.; Tsunoda, K.I.; Wang, J. A solid-cladding/liquid-core/liquid-cladding sandwich optical waveguide for the study of dynamic extraction of dye by ionic liquid BmimPF 6. *Appl. Spectrosc.* **2012**, *66*, 798–802.

60. Shi, Y.; Liang, L.; Zhu, X.; Zhang, X.; Yang, Y. Tunable self-imaging effect using hybrid optofluidic waveguides. *Lab Chip* **2015**, *15*, 4398–4403.

61. Choi, J.; Lee, K.S.; Jung, J.H.; Sung, H.J.; Kim, S.S. Integrated real-time optofluidic SERS via a liquid-core/liquid-cladding waveguide. *RSC Adv.* **2015**, *5*, 922–927.

62. Lim, J.M.; Kim, S.H.; Choi, J.H.; Yang, S.M. Fluorescent liquid-core/air-cladding waveguides towards integrated optofluidic light sources. *Lab Chip* **2008**, *8*, 1580–1585.

63. Yang, Y.; Liu, A.Q.; Lei, L.; Chin, L.K.; Ohl, C.D.; Wang, Q.J.; Yoon, H.S. A tunable 3D optofluidic waveguide dye laser via two centrifugal Dean flow streams. *Lab Chip* **2011**, *11*, 3182–3187.

64. Yang, Y.; Liu, A.Q.; Chin, L.K.; Zhang, X.M.; Tsai, D.P.; Lin, C.L.; Lu, C.; Wang, G.P.; Zheludev, N.I. Optofluidic waveguide as a transformation optics device for lightwave bending and manipulation. *Nat. Commun.* **2012**, *3*, 651.

65. Watts, B.R. Development of a Microchip-Based Flow Cytometer with Integrated Optics—Device Design, Fabrication , and Testing. Ph.D Thesis, McMaster University, Hamilton, ON, Canada, 2012.

66. Emile, O.; Emile, J. Soap films as 1D waveguides. *Optofluid. Microfluid Nanofluid.* **2014**, *1*, 27–33.

67. Watts, B.R.; Kowpak, T.; Zhang, Z.; Xu, C.Q.; Zhu, S. Formation and characterization of an ideal excitation beam geometry in an optofluidic device. *Biomed. Opt. Express* **2010**, *1*, 848–860.

68. Watts, B.R.; Zhang, Z.; Xu, C.Q.; Cao, X.; Lin, M. A photonic-microfluidic integrated device for reliable fluorescence detection and counting. *Electrophoresis* **2012**, *33*, 3236–3244.

69. Watts, B.R.; Zhang, Z.; Xu, C.Q.; Cao, X.; Lin, M. Integration of optical components on-chip for scattering and fluorescence detection in an optofluidic device. *Biomed. Opt. Express* **2012**, *3*, 2784–2793.

70. Watts, B.R.; Zhang, Z.; Xu, C.Q.; Cao, X.; Lin, M. A method for detecting forward scattering signals on-chip with a photonic-microfluidic integrated device. *Biomed. Opt. Express* **2013**, *4*, 1051–1060.

71. Ungerbock, B.; Charwat, V.; Ertl, P.; Mayr, T. Microfluidic oxygen imaging using integrated optical sensor layers and a color camera. *Lab Chip* **2013**, *13*, 1593–1601.
72. Yang, S.; Undar, A.; Zahn, J.D. A microfluidic device for continuous, real time blood plasma separation. *Lab Chip* **2006**, *6*, 871–880.
73. Eyer, K.; Root, K.; Robinson, T.; Dittrich, P.S. A simple low-cost method to enhance luminescence and fluorescence signals in PDMS-based microfluidic devices. *RSC Adv.* **2015**, *5*, 12511–12516.
74. Kettlitz, S.W.; Valouch, S.; Sittel, W.; Lemmer, U. Flexible planar microfluidic chip employing a light emitting diode and a PIN-photodiode for portable flow cytometers. *Lab Chip* **2012**, *12*, 197–203.
75. Hoera, C.; Ohla, S.; Shu, Z.; Beckert, E.; Nagl, S.; Belder, D. An integrated microfluidic chip enabling control and spatially resolved monitoring of temperature in micro flow reactors. *Anal. Bioanal. Chem.* **2014**, *407*, 387–396.
76. Nedbal, J.; Visitkul, V.; Ortiz-Zapater, E.; Weitsman, G.; Chana, P.; Matthews, D.R.; Ng, T.; Ameer-Beg, S.M. Time-domain microfluidic fluorescence lifetime flow cytometry for high-throughput Förster resonance energy transfer screening. *Cytom. A* **2015**, *87*, 104–118.
77. Liu, H.; Crooks, R.M. Three-dimensional paper microfluidic devices assembled using the principles of origami. *J. Am. Chem. Soc.* **2011**, *133*, 17564–17566.
78. Nge, P.N.; Rogers, C.I.; Woolley, A.T. Advances in microfluidic materials, functions, integration, and applications. *Chem. Rev.* **2013**, *113*, 2550–2583.
79. Zhang, Z.; Zhao, P.; Xiao, G.; Watts, B.R.; Xu, C. Sealing SU-8 microfluidic channels using PDMS. *Biomicrofluidics* **2011**, *5*, 1–8.
80. Ren, Y.; Huang, S.H.; Mosser, S.; Heuschkel, M.O.; Bertsch, A.; Fraering, P.C.; Chen, J.J.J.; Renaud, P. A Simple and Reliable PDMS and SU-8 Irreversible Bonding Method and Its Application on a Microfluidic-MEA Device for Neuroscience Research. *Micromachines* **2015**, *6*, 1923–1934.
81. Yang, C.C.; Wen, R.C.; Shen, C.R.; Yao, D.J. Using a microfluidic gradient generator to characterize BG-11 medium for the growth of cyanobacteria synechococcus elongatus PCC7942. *Micromachines* **2015**, *6*, 1755–1767.
82. Chen, H.T.; Wang, Y.N. Optical microflow cytometer for particle counting, sizing and fluorescence detection. *Microfluid. Nanofluid.* **2009**, *6*, 529–537.

micromachines

MDPI

Review

Liquid Core ARROW Waveguides: A Promising Photonic Structure for Integrated Optofluidic Microsensors

Genni Testa *, Gianluca Persichetti and Romeo Bernini

Istituto per il Rilevamento Elettromagnetico dell'Ambiente, Consiglio Nazionale delle Ricerche (IREA-CNR), Via Diocleziano 328, 80124 Naples, Italy; persichetti.g@irea.cnr.it (G.P.); bernini.r@irea.cnr.it (R.B.)
* Correspondence: testa.g@irea.cnr.it; Tel.: +39-08-1762-0643; Fax: +39-08-1570-5734

Academic Editors: Shih-Kang Fan, Da-Jeng Yao and Yi-Chung Tung
Received: 16 January 2016; Accepted: 7 March 2016; Published: 11 March 2016

Abstract: In this paper, we introduce a liquid core antiresonant reflecting optical waveguide (ARROW) as a novel optofluidic device that can be used to create innovative and highly functional microsensors. Liquid core ARROWs, with their dual ability to guide the light and the fluids in the same microchannel, have shown great potential as an optofluidic tool for quantitative spectroscopic analysis. ARROWs feature a planar architecture and, hence, are particularly attractive for chip scale integrated system. Step by step, several improvements have been made in recent years towards the implementation of these waveguides in a complete on-chip system for highly-sensitive detection down to the single molecule level. We review applications of liquid ARROWs for fluids sensing and discuss recent results and trends in the developments and applications of liquid ARROW in biomedical and biochemical research. The results outlined show that the strong light matter interaction occurring in the optofluidic channel of an ARROW and the versatility offered by the fabrication methods makes these waveguides a very promising building block for optofluidic sensor development.

Keywords: liquid ARROW; optofluidics; microfluidics; optical sensors; integrated silicon technology

---

1. Introduction

Optofluidics is an emerging research field that combines the advantages of microfluidics and optics on the same platform towards highly functional and compact devices [1,2]. More precisely, optofluidic approaches provide high-yield integration of fluidics by including fluidic elements as parts of the photonic structure. This solution confers to the optofluidic systems a high level of compactness and reconfigurability as the fluids provide an effective means to tune and reconfigure the photonic architecture.

In the field of optical sensor technologies, optofluidics offers innovative design solutions in order to integrate all the required microfluidic and optical functionalities on a single chip for high-throughput analysis. Optofluidic sensors have been extensively developed for biomedical and biochemical analysis for healthcare purposes (*i.e.*, diagnostic), pharmaceutical research, and environmental monitoring [3]. Whatever optical method is used, highly-sensitive detection requires a very careful design of both optical and microfluidic parts in order to push the device sensitivity to the limit of the theoretical capability. Furthermore, innovative design strategies need to be developed to overcome the reduced interaction length owing to devices size miniaturization. In this context, optofluidics answer to the growing need for miniaturized and portable sensing devices, highly integrated, and extremely sensitive [4].

Optofluidic waveguides represent one of the simplest examples of optofluidic integration, as the optical and fluidic structures are tightly integrated into each other. In particular, an optofluidic waveguide is a photonic structure able to perform optical confinement of the light through a fluid [5]. Among others, liquid cores antiresonant reflecting optical waveguides (ARROWs) have been revealed as a very powerful building-block for optofluidic sensors [6]. In an ARROW the light is guided through the liquid core by means of highly reflective mirrors suitably realized on the core sidewalls. Since both the light and the liquid are guided through the same microchannel, ultra-high sensitivity can be achieved by exploiting the direct interaction of the entire optical waveguide mode with the fluid samples. Such a strong optical coupling, combined with the ultra-small liquid volume, enables sensitivity down to single molecule level, which is the ultimate goal of analytical methods. ARROWs can be easily fabricated using silicon technologies; moreover, like others optofluidic waveguides, such as the photonic crystal [7,8] and the slot waveguides [9,10], ARROWs are, in essence, planar and, hence, highly attractive for planar optofluidic integration [11,12]. As a matter of fact, starting from the first demonstration of the ARROWs' potentialities in the field of optical sensing, there has been a growing interest for them as basic elements in optofluidic chip for sensing applications.

In the following, we report on optofluidic sensors realized with liquid ARROWs. Starting from the description of the working principle and fabrication methods of liquid ARROWs, we review the state of the art of on chip integration for sensing applications, highlighting specific advantages gained through the optofluidic integration.

2. Operation Principles and Simulations

2.1. 1-D Liquid ARROW

ARROW waveguides enable light confinement and propagation thanks to interference phenomena occurring in the cladding structure, made up of alternating high- and low- refractive index (RI) thin layers. The working principle of an ARROW waveguide can be easier understood by considering a schematic one-dimensional structure, as shown in Figure 1. For simplicity, we consider only two cladding layers with refractive index n_1 and n_2 ($n_1 > n_2$). Light propagates through the core by means of Fresnel reflections at the cladding interfaces. The entrapped light in each cladding layer experiences interference due to repeated reflections at cladding interfaces. In order to intensify the light reflected back into the core, each interference cladding must be designed to fulfill a specific condition. By considering the analogy of a cladding layer with a Fabry–Pérot cavity, such condition corresponds to the anti-resonance state of the cavity, which performs the maximum reflectivity. The reflectivity of the multilayer stack constituted by multiple interference claddings, generally called Fabry–Pérot mirror, can be very high, up to 99%.

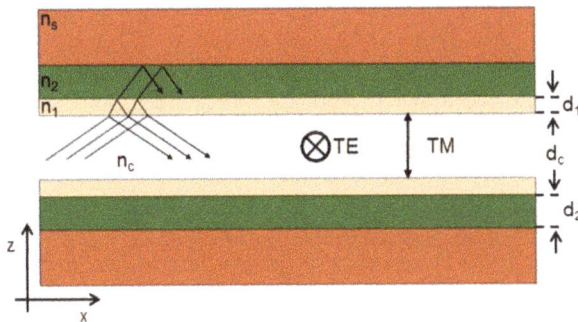

Figure 1. 1-D structure of ARROW waveguide.

A detailed description of the ARROW confinement phenomenon is reported in [13], where the phase accumulated by the fundamental mode at each round trip in the interference claddings is clearly defined. By setting the specific condition of antiresonance, a very useful expression to calculate the optimal cladding thicknesses d_1 and d_2, can be derived:

$$d_{1,2} = \frac{\lambda}{4} \left[n_{1,2}^2 - n_{eff}^2 \right]^{\frac{-1}{2}} (2N+1) \qquad N = 0, 1, 2, \ldots \tag{1}$$

where λ is the working wavelength, $n_{1,2}$ is the refractive index of the cladding layers and n_{eff} is the effective index of the propagation mode. For the fundamental mode, n_{eff} can be approximately given by:

$$n_{eff} = \frac{\beta}{\beta_0} \cong n_c \left[1 - \left(\frac{\lambda}{2 n_c d_c} \right)^2 \right]^{\frac{1}{2}} \tag{2}$$

where $\beta_0 = 2\pi/\lambda$, β is the propagation constant, n_c and d_c are the refractive index and the thickness of the core, respectively. If $\lambda/d_c << 1$ (e.g., large core waveguides), Equation (2) can be simplified to $n_{eff} \approx n_c$. As long as the operating wavelength and the core and cladding materials are conveniently chosen, reduced propagation losses can be obtained by using Equations (1) and (2) to calculate the cladding thickness.

It should be noted that, unlike photonic crystal waveguides, in ARROW waveguides the high- and low-index alternating layers do not require a high periodicity in the structure [14]. This characteristic simplifies the waveguides manufacturing and strongly relaxes the fabrication constrains.

ARROWs are leaky waveguides and require a very careful design to operate in a single-mode condition, which is a fundamental requirement for sensing applications, *i.e.*, for the fabrication of interferometric devices. Single-mode operation can be accomplished by exploiting the weaker reflectivity experienced by higher order modes upon reflections at cladding interfaces. Low-order modes propagate with glancing angles at cladding interfaces, hence the Fresnel coefficient is quite high and the loss is low, whereas higher-order modes have larger incident angles and hence experience higher losses. An extensive investigation of the ARROW attenuation losses has been reported in [15]. The authors provide very useful analytical expressions that, with a good approximation, describe the mode losses for both TE and TM-polarized modes. A simple calculation form analytical expression given in [15] permits to calculate the ARROW attenuation coefficient for the TE and TM fundamental modes in the following, very useful, approximated form:

$$\begin{cases} \alpha_{TE} = \dfrac{0.5\lambda^4 \left[1 + \dfrac{4d_c^2}{\lambda^2} \left(n_2^2 - n_c^2 \right) \right]}{n_c d_c^5 \left(n_1^2 - n_c^2 \right) \sqrt{\left(n_s^2 - n_c^2 \right)}} \\[2em] \alpha_{TM} = \left(\dfrac{n_1^2 n_s}{n_c n_2^2} \right)^2 \alpha_{TE} \end{cases} \tag{3}$$

which is valid when $d_c >> \lambda$; n_s is the refractive index of the substrate. By assuming Si_3N_4 ($n_1 = 2.01$) and SiO_2 ($n_2 = 1.44$) for the first and second layer, Si ($n_s = 3.8$) for the substrate, and water core ($n_c = 1.33$), the proportional coefficient between α_{TM} and α_{TE} in Equation (3) is as high as 30. Hence the propagating modes experience attenuation coefficients that strongly depend upon their state of polarization. In particular, propagation of TE-polarized light is always privileged due to the lower attenuation coefficients. Additionally, in this case this behavior originates from the light confinement mechanism, based on Fresnel reflections at core and cladding boundaries. Fresnel reflection depends on the polarization of the incident light and TM reflection coefficient is always lower with respect to TE reflection.

Moreover, as a general rule, from Equation (3) the attenuation coefficients decrease by increasing the core dimensions. This implies that the core size must be suitably chosen in order to reduce fundamental mode loss but also to minimize the contribution of higher order modes to the total transmitted power. The degree of suppression of these modes improves as the core width decreases [16]. A further very useful general rule can be retrieved: in order to reduce the attenuation coefficients, the refractive index of the first cladding (n_1) should be higher than the core index, while the refractive index of the second cladding (n_2) as closer as possible to the core index (n_c). These requirements impose very important limitations to the choice of the cladding materials. For this reason typical materials are: silicon nitride (n_1 = 2.01–2.2), titanium dioxide (n_1 = 2.49), and tantalum dioxide (n_1 = 2.3) for the first high index cladding and silicon dioxide (n_2 = 1.46) for low index cladding, as they provide proper index contrast, for water core waveguides, and simple fabrication by means of silicon compatible process.

By combining these intrinsic properties along with an accurate design strategy, ultra-low loss, single mode linearly polarized ARROWs operation can be accomplished and exploited to realize high-performance interferometric devices.

2.2. 2-D Liquid ARROW

In a practical ARROW waveguide, the core must be completely surrounded by the cladding layers to accomplish light-confinement in both lateral and vertical direction. One possible ARROW structure for both vertical and horizontal confinement is illustrated in Figure 2.

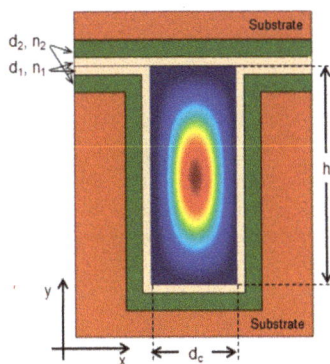

Figure 2. Cross-section of a 2-D liquid ARROW waveguide with simulated power distribution of the fundamental mode.

A simple estimation of the attenuation coefficients in 2-D ARROW waveguides can be performed by using the one dimensional analytical model [11,15]. In a paraxial approximation the problem of the 2-D ARROW confinement can be separated into two 1-D problem for the lateral and transverse direction, in this way the evaluation of the total losses can be obtained from the addition of the resulting 1-D losses. In Figure 3 the losses for the fundamental (blue line) and first order mode (red line) *versus* wavelength calculated by using approximated analytical model, are shown. Hence, as anticipated, for a fixed core dimension higher order modes attenuate faster than the fundamental mode. From the figure the typical transmitted spectrum of an ARROW waveguide can also be foreseen. It is characterized by a broad spectral band, corresponding to the range of wavelengths which experience destructive interference in the claddings. The wavelength dependence of the propagation losses could be used to tailor the ARROW spectral response in sensing applications, *i.e.*, to filter out unwanted optical wavelengths from reaching the detector.

In 2-D ARROW design, the influence of the input polarization on waveguide losses should be taken carefully into account. If incident light is polarized along the *y*-axis (Figure 2) it is reflected upon lateral cladding as TE-polarized light whereas it is TM-polarized with respect to the vertical ARROW confinement. Hence, TE-polarized input light along *x*-axis will also contribute to the total losses as a TM polarization with respect to *y*-axis. By considering the polarization dependence of ARROW losses, these considerations suggest that the polarization of the propagating modes can be selected by suitably designing the core shape. If one of the core dimension is set wider (*i.e.*, a rectangular core shape), modes polarized along the longer side experience lower TM propagation losses with respect to the modes polarized along the crossing direction, *i.e.*, the longer core side identifies the direction of the supported polarization.

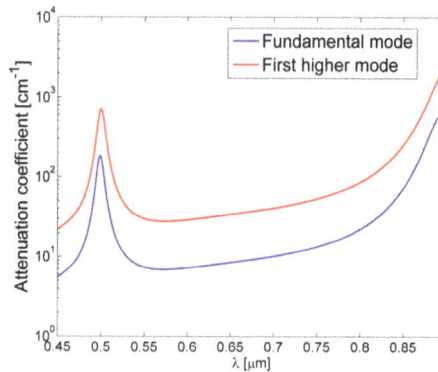

Figure 3. Attenuation coefficient of the fundamental mode ($E_y{}^0$) and the first order mode ($E_y{}^{12}$) *versus* the wavelength in water-core ARROW waveguide with d_1 = 290 nm, d_2 = 260 nm, d_c = 5 μm and h = 10 μm (Figure 2).

More accurate loss evaluation requires 2-D modes numerical analysis. A first detailed numerical investigation of 2-D hollow core ARROWs for sensing application is reported in [6]. The authors simulated ARROWs propagation by using finite element method (FEM) numerical analysis. The propagation characteristics of an ARROW with a hexagonal core cross section have been studied for different diameters of the core, filled with air or liquid substances. The authors also investigated the attenuation losses characteristics of the fundamental mode *versus* the detuning of the thickness of the first and second cladding layer from its antiresonant value. As it was found for solid ARROW [13], their study confirms that the attenuation losses are almost unaffected by the second layer thickness changes in a range of ±30 nm, whereas it is very sensitive to first cladding thickness, increasing very rapidly for thickness detuning of about ±10 nm. These results provide very useful rules for waveguides fabrication. 2-D numerical analysis of ARROWs modes characteristics have been also performed by using finite difference methods (FDM) [16,17].

3. Fabrication and Characterization

Currently, ARROWs are fabricated with two main different approaches: by bulk micromachining process or by surface micromachining process. In the first case, the ARROW is constituted by two halves of silicon substrate with the core etched in one of the halve part. The two processed silicon substrates are finally bonded after cladding depositions on both halves [17,18]. In the second case, the core is obtained by depositing and patterning a sacrificial material on a single silicon substrate in between top and bottom claddings. The sacrificial material is removed by etching at the end of process [19,20]. In both processes cladding layers are deposited using deposition techniques like plasma enhanced chemical vapor deposition (PECVD), low pressure chemical vapor deposition (LPCVD) or atomic layer deposition (ALD).

Both fabrication processes have some advantages and disadvantages. Bulk micromachining is a faster process but it needs a non-standard wafer bonding between high index materials like silicon nitride or titanium dioxide. However it enables a flexible design, as the top substrate can be substituted, allowing for the implementation of hybrid configuration, as it will be illustrated in the following. On the other hand, sacrificial etching of long narrow micro channels is very time consuming and it requires controlled manufacturing in order to achieve 100% fabrication yields [20]. Nevertheless, this approach allows a simple monolithic integration with solid core exciting/collecting waveguides. As it will be showed in the following, in both cases ARROW waveguides have been successfully produced and applied to fabricate very high performance and promising optofluidic devices for sensing applications.

Single-mode ARROWs with hollow core have been firstly demonstrated by Yin *et al.* [21]. Hollow core dimensions were 3.5 μm by 12 μm, obtained by sacrificial etching of SU-8 (Figure 4a). For low-loss propagation, four antiresonant claddings were fabricated by alternating silicon oxide and nitride layer depositions on a silicon substrate using PECVD. The waveguide was designed to operate at the wavelength of 785 nm and with $n_c = 1$. The authors demonstrated single-mode operation with propagation loss as low as 6.5 cm^{-1}. The same waveguide was also characterized with a liquid-filled core ($n_c = 1.33$), showing waveguide attenuation coefficient of 2.4 cm^{-1} at $\lambda = 633$ nm [19]. Successive improvements in the fabrication process have been proposed by Yin *et al.* [22] to reduce the waveguide losses as low as 2.6 cm^{-1} for 15×3.5 μm^2 core dimensions ($n_c = 1$). In this work the authors also discussed the effect on the waveguides losses of different growing rate over the horizontal and vertical core walls, which is intrinsic in the PECVD process. This effect causes that the vertical and horizontal layers do not simultaneously fulfill the antiresonance condition, deteriorating the waveguide performances. Even if this effect can be compensated by appropriate layer design [23], it can lead to non-optimal waveguide losses. In the work of Testa *et al.* [24] the ALD process has been proposed to overcome these limitations. ARROW waveguide was fabricated by using bulk micromachining, starting with the realization of the microchannels by etching 5 μm in the silicon wafer. Alternating layer of silicon dioxide and titanium dioxide were deposited on the core sidewalls by using LPCVD and ALD, respectively. One of the advantages of the ALD technique is the possibility to deposit very thin cladding layer (sub-100 nm) with an excellent conformality, reproducibility, and with a high precision over the resulting thickness. The waveguides were designed to operate on single mode at $\lambda = 635$ nm with a water filled core ($n_c = 1.33$). The fundamental mode loss of 5.22 cm^{-1} was obtained with only two antiresonant layers. In Figure 5 the measured and simulated transmitted spectra of a 1.5 cm long ARROW waveguide fabricated by ALD are shown. Simulations have been performed by using by using a commercial software (FIMMWAVE, Version 5.2, © Photon Design) and FDM method. In the simulation model we have taken into account the wavelength dependence of the materials composing the waveguiding structure and the liquid filling the core (methanol). As Figure 5 shows, there is a discrepancy between the simulated and measured spectrum at low wavelength (below 400 nm). This discrepancy is due to the difficulty to obtain an accurate model for the core and the cladding layers refractive indexes (both real and imaginary parts) at wavelengths lower than 400 nm. In particular, as it can be noticed in figure, the spectral region at around 370 nm corresponds to the resonance condition in the cladding layers. As it is demonstrated in the [25], the wavelength position of the minimum in the transmitted spectrum (resonance in the claddings) is greatly influenced by the core and cladding refractive indexes.

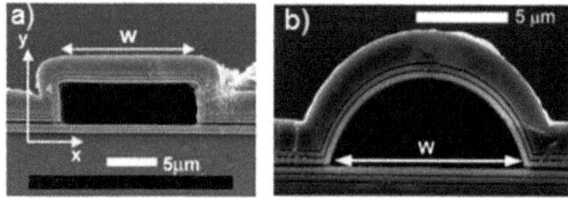

Figure 4. Scanning electron microscope (SEM) images of (**a**) hollow-core ARROWs with rectangular and (**b**) arch-shaped cross sections fabricated by surface micromachining process. Reprinted with permission from [27]. Copyright (2005) OSA.

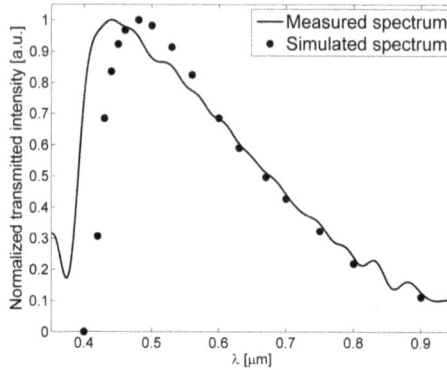

Figure 5. Measured and simulated transmitted spectrum from a single mode ARROW waveguide fabricated by ALD with d_1 = 75.4 nm (TiO$_2$), d_2 = 262 nm (SiO$_2$), d_c = 10 μm, h = 5 μm, n_c = 1.32 (methanol-filled core).

Single-mode ARROWs with arc-shaped core were also demonstrated [26] (Figure 4b). The authors fabricated such structures by sacrificial etching and demonstrated their ability to overcome the mentioned limitation of the PECVD process. Further important advantages compared with rectangular ARROWs are the higher mechanical stability and the smoother core walls. Optical characterization of such structures with four antiresonant claddings was also carried out [27]. A low attenuation coefficient of 0.26 cm^{-1} was obtained at λ = 633 nm with a high refractive index liquid core (ethylene glycol, n_c = 1.43).

Single mode hybrid silicon-polymer ARROW (h-ARROW) waveguide has been achieved by Testa *et al.* [28]. The authors exploited the advantage of bulk micromachining to separately process two silicon wafers, which were then bonded at the end of the cladding deposition. H-ARROW was obtained by substituting the upper silicon part with a single polydimethylsiloxane (PDMS) layer. Slightly increased optical losses were expected and measured due to the weaker optical confinement at PDMS-core interface compared with the stronger confinement enabled by antiresonant claddings. However, thanks to a proper design, this effect was minimized and fundamental mode loss of 6.18 cm^{-1} at λ = 633 nm (n_c = 1.33) was obtained. A detailed study of h-ARROW confinement and loss improvement is reported in [16]. This approach circumvents the need of non-standard wafer bonding step between high index materials like silicon nitride or titanium dioxide, which is substituted by a very simple temporary or permanent bonding between silicon nitride and PDMS.

Single-mode liquid ARROWs with high performance have been demonstrated by using the two above-mentioned micromachining processes. Surface micromachining is very promising as it enables monolithic integration with exciting and collecting waveguides. As it can be envisioned, other components can be fabricated and integrated on the same platform in order to add further interesting functionalities for complete lab-on-a-chip realization. Bulk micromachining promises very high flexibility enabled by hybrid approaches. Polymer materials exhibit very favorable properties for both photonic and microfluidic technologies; moreover, thanks to the related inexpensive fabrication procedures, they show a great potential for economic mass production of lab-on-a-chip devices.

4. ARROW-Based Devices and Applications

Optofluidic waveguides have been demonstrated as very promising candidates for high performance optical microsensors [3,5]. Liquid ARROWs play a prominent role in this context, enabling planar integrated fabrication approaches based on the currently available and well-established silicon compatible technology.

Currently, the optical detection methods mainly applied in optical microsensors includes spectroscopic techniques like absorption, fluorescence, and Raman scattering. These approaches have demonstrated to be very effective for highly selective and sensitive detection of molecules down to the single molecule level. Additionally, the detection based on refractive index changes is widely used and have demonstrated a great potential for label-free detection of biomolecules with very high sensitivity.

In the following, we describe optofluidic design approaches for optical microsensors realized with liquid ARROWs.

4.1. Liquid ARROW Waveguides as Optofluidic Sensors

The strong light matter interaction occurring in the liquid core of an ARROW waveguide has been exploited to realize simple but effective and sensitive microfluidic sensors.

In the work of Campopiano *et al.* [25] the first demonstration of an integrated optical bulk refractometer based on multimode liquid ARROW has been proposed. For fluid injection, inlet, and outlet openings were fabricated in the silicon substrate. A multimode ARROW with large core section of 130×130 μm^2 was used. The exciting and collecting optical fibers were inserted directly into the core, resulting in a highly compact and robust self-aligned configuration. Very notably, the waveguide, itself, constitutes the sensor. The sensing principle was based on the intrinsic property of the ARROW: the shift of the minimum of the transmitted spectrum upon core refractive index changes. Good bulk sensitivity of 555 nm/RIU (RIU: refractive index unit) and linear response, with a limit of detection (LOD) of 9×10^{-4} RIU has been demonstrated. The same configuration has been also used to realize an integrated long path absorbance cell for colorimetric detection of specific protein in water solution [17]. The authors used a Bradford assay for colorimetric determination of bovine serum albumin, showing an LOD of about 1 μg/mL with a sample volume of only 0.34 nL and a 15 mm-long-cell. This example was the first experimental demonstration of the sensing capability of a simple straight liquid core ARROW.

Multimode liquid ARROWs have also been employed to realize a flow cytometer for the analysis of fluorescently-labeled human T leukemia cells (Jurkat) [29]. The excitation light for cells interrogation propagated in the optofluidic channel thanks to the ARROW confinement, resulting in a highly compact and efficient optical interrogation scheme. The cells stream was hydrodynamically focused in the center of the channel where two optical fibers were arranged orthogonally for collecting the emitted fluorescent signals.

However, multimode ARROW with large core section area is not preferred to reach ultra-high sensitivity due to the large sample volume that typically causes strong background signal. This objective has been met by using single mode liquid ARROW. Thanks to the core section of a few microns, they provide a very small excitation volume, which is attractive for detection down to single molecule sensitivity.

In the work of Yin *et al.* [23], a liquid ARROW with picoliter core volume has been optimized and applied for highly efficient fluorescence detection of dye molecules, demonstrating an LOD corresponding to 490 molecules in the core. The potential for single molecule sensitivity by surface-enhanced Raman scattering (SERS) detection has been also demonstrated by Measor *et al.* [30] using a simple straight single mode ARROW.

The sensors above described can be considered as the first generation of ARROW-based optofluidic chips. However, more advances can originate from the use of optofluidic approaches to realize multifunctional photonic device and microfluidic elements on the same chip, as it will be illustrated in the following.

4.2. Interferometric Optofluidic Devices for Sensing Applications

Multimode and single-mode liquid ARROWs have been demonstrated as very promising tools for sensing applications because of their high compactness, sensitivity, and the inherit ability to transport the light and the analyte in the same physical channel. The potential for the realization of various functional optical elements on chip makes ARROW a suitable candidate for integration into compact photonic devices. Additionally, simply using the waveguide itself as optofluidic sensor, more sophisticated device architectures can, in principle, be realized by using liquid core ARROWs. In particular, the demonstration of single-mode ARROWs has provided the opportunity to implement optical sensors which make use of interferometric phenomena. For instance multimode interference (MMI) splitters based on liquid ARROWs have been demonstrated [31] and implemented in on-chip optofluidic interferometers [32]. Moreover, curved ARROW sections can be used for the realization of optical devices comprising light splitters or combiners, as it occurs in integrated optical interferometers. Bend waveguides based on ARROWs have been modeled by using 1-D FDM method [16] and fabricated in ring resonator geometry [33].

Planar optical integrated interferometric devices are very attractive for sensing applications as they combine high sensitivity with compactness. Typically, these devices are based on conventional solid core waveguides and exploit evanescent sensing to detect the presence of analytes in fluid sample [34–36]. Recently, innovative optofluidic architectures have been proposed for these devices to simplify fluids sampling and manipulation via microfluidic integration [4]. Very interesting optofluidic architectures for interferometric devices have been proposed based on droplet microfluidics [37] and fluid-air interface (meniscus) [38,39]. Several integrated optofluidic interferometric devices have been demonstrated based on liquid ARROWs, like wavelength division multiplexing (WDM) using MMI [40], ring resonators [41], and Mach–Zehnder interferometers (MZIs) [32,42].

Optofluidic schemes for integrated MZIs in which a microfluidic channel is inserted along the sensing arm have been fabricated to allow the direct coupling of the light with sample fluid [43,44]. This design strategy enables a higher sensitivity thanks to the almost complete overlap of the mode power with the sample fluid in the sensing region. However, in these cases the interaction length is limited by the width of the microfluidic channel. Moreover, the high insertion losses at solid/liquid interface in the inserting region can cause poor device performances. With the advantages offered by liquid ARROW, a planar integrated optofluidic MZI entirely shaped by liquid microchannels has been fabricated [42]. The design consists of two asymmetric single-mode branches split from a Y-Branch and recombined by an inverse Y-Branch in single mode waveguide for interfering. Further improvement in sensor design has been proposed in the work of Testa *et al.* [32] to minimize the power unbalancing between arms and to increase the visibility of the interferometer (Figure 6a). The device features large RI tunability, typical of the optofluidic microsystems, as the guiding condition is satisfied for quite a large RI range $n_c = 1.32 \div 1.45$, that corresponds to the detection range of the sensing device. RI limit of detection (LOD) of $\Delta n = 1.6 \times 10^{-5}$ RIU has been estimated, which compares well with results reported for other optofluidic MZI for label-free sensing of liquids [43,45]. However, while evanescent based-MZIs have been demonstrated as extremely sensitive in direct and real-time measurement of biomolecular interactions [46], optofluidic MZIs have been mostly applied for RI volume sensing of liquids due to the enhanced bulk interaction occurring in the optofluidic channel. Detection of single viruses and the discrimination between different kinds of virus types in a sample mixture have been demonstrated using an optofluidic heterodyne interferometer employing an out-of-plane detection/interrogation optical scheme [47].

Optofluidic ring resonators (ORRs) have also been recently proposed. Compared with MZI-based sensors that exploit limited sensing lengths, optical resonators benefit from the repetitive interaction between the light and the sample, achieved thanks to the resonant recirculation of light in the cavity. Thanks to this property, highly compact devices can be fabricated with very high sensing capability. Several architectures for optofluidic resonators have been demonstrated which incorporate fluidic capabilities into the photonic structure, such as the capillary-based optofluidic ring resonators (CORRs) [48–50]. The CORR is a microfluidic capillary that supports resonant optical modes (the so-called whispering gallery modes (WGMs)) circulating in the capillary wall with the evanescent field extending into the fluid.

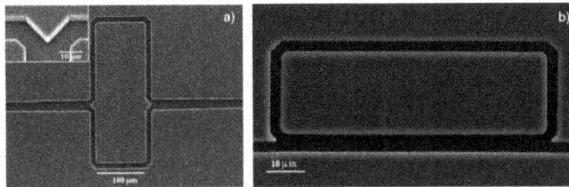

Figure 6. (a) SEM image of an integrated MZI based on liquid ARROWs. Reprinted with permission from [32]. Copyright (2010) OSA. (b) SEM image of an integrated optofluidic ring resonator based on liquid ARROWs. Reprinted with permission from [41]. Copyright (2010) AIP.

Liquid core ARROW has been successfully employed as a basic element to form an integrated ORR (Figure 6b) [41]. A first generation of ARROW-based ORR has been fabricated with a moderate quality factor (Q-factor) on the order of 10^3, mainly due to the high optical losses of the curved sections. Advances in the ORR performances were experimentally obtained by performing an accurate optimization on the initial structure design (first-generation ORR), including those concerning the optofluidic level of integration [16]. A planar ORR based on liquid ARROWs with a Q-factor of about 10^4 has been demonstrated by applying the optimization procedure [33]. The proposed approach retained the merit of high level of optofluidic integration in a planar configuration. The strong light-matter interaction occurring in the liquid core confers high bulk RI sensitivity to the ORR, experimentally estimated as about 700 nm/RIU. This value is larger than that obtained by most of the other optofluidic ring resonator sensors like slot- and capillary-based ORRs [50–52]. In order to be applied for biomolecules detection, the surface sensing performances of the device have been simulated, showing very promising capabilities [16]. In comparison, the CORRs have demonstrated very good surface sensing. Biotin molecules captured on the wall surface have been detected with a detection limit of approximately 1.6 pg/mm^2 [51]. Breast cancer biomarkers (CA15-3) have been also detected at a concentration of 1–200 units/mL in phosphate-buffered saline buffer and 20–2500 units/mL in fetal calf serum [53]. However, tapered optical fibers are typically used to excite WGMs, which hinders the integration of CORRs onto a compact lab-on-chip platform.

4.3. Planar Integrated ARROW Platform

One of the most relevant problems associated with the chip-scale integration of liquid core waveguides concerns the way light is collected from the open-ended liquid core. An optimized transmission between liquid core waveguides and off-chip components is necessary to increase the overall optical throughput. In particular, the ability to efficiently collect the light is crucial for achieving single-particle detection. Not surprisingly, the first demonstration of single molecule sensitivity using an open-ended liquid ARROW required bulk optical components, such as a high numerical aperture objective lens [23,30].

Technological solutions to provide on chip integration of intersecting microfluidic channels and optical waveguides are an object of many works in the literature [54–58]. In addition to the obvious implications to enable miniaturization and simplify the alignment procedure, on-chip integrated waveguides offer two further remarkable advantages. First, the possibility to improve the optical coupling at the waveguide-fluid interface that in turn improves the signal to noise ratio and the sensitivity of detection. Second, they can be suitably placed on-chip in order to illuminate only specific points along the liquid core waveguide, thus enabling highly spatially-resolved optical detection on a sub-picoliter excitation volume. Using surface micromachining technology, the first step towards the realization of an integrated platform based on liquid ARROW has been proposed in the work of Yin *et al.* [59]. The authors demonstrated an important advance in sensor design, obtained by integrating solid core with liquid core ARROW in a full planar geometry. Parallel arrays of solid core waveguides were arranged orthogonally with liquid waveguides for fluorescence measurements on sub-picoliter volumes. The proposed approaches resulted in a network of intersecting waveguides for specific excitation of well-defined locations along the liquid core waveguide. Thanks to this geometry, strong pump suppression and sensor sensitivity down to a single molecule have been achieved. Using a similar approach, a solid core waveguide aligned with the liquid waveguide has been integrated on an ARROW platform for surface-enhanced Raman spectroscopy (SERS) detection with molecular specificity [60]. In this configuration the solid waveguide was collinear with the liquid waveguide and it was used to excite the entire liquid volume. Moreover, in order to prevent liquid evaporation during measurements, a further remarkable improvement has been proposed, consisting in an on-chip reservoir of 10 µL volume for continuous piping of fluid solution. This has been the first implementation of a reliable optofluidic tool on an ARROW sensing platform. Using this platform, the authors detected the minimum active R6G concentration of 30 nM, corresponding to measuring approximately 4×10^5 molecules in the ARROW volume of 44 pL. Other very interesting optofluidic devices for SERS detection have been demonstrated with better limits of detections, but generally requiring out-of-plane bulk instruments for the measurements [61,62].

Step-by-step, successive improvements of the optical and sensing performances of the liquid ARROWs have also been made from a fabrication point of view [63–66]. In [65] tantalum oxide and low temperature-deposited silicon nitride were proposed to reduce the background contribution of native photoluminescence of the cladding material in fluorescence measurements. With this arrangement the authors demonstrated a signal-to-noise-ratio (SNR) improvement by a factor of 12 in fluorescent nanoparticle detection compared to conventional silicon nitride-based ARROWs. Optimization of interface transmission between the liquid and solid ARROWs has also been performed to achieve high sensitivity in spectroscopic measurements [66,67].

A chip-scale integrated planar sensing system that combines optical and fluidic functions is required to address the challenge of high sensitivity and compactness. By using a liquid ARROW integrated with both orthogonal and aligned solid waveguides, together with sample reservoirs, the first planar optofluidic chip for single particle detection and analysis has been demonstrated [68]. The chip has been fabricated by using surface micromachining planar silicon technology. In this design, solid waveguides are not only used to precisely excite a sub-picoliter liquid volume but also to collect and deliver light from liquid ARROW towards off-chip fibers (Figure 7). The advantages offered by this design of providing multiple and spatially separated collection paths have enabled the implementation of advanced multi-color spectroscopy techniques on chip, such as fluorescence cross-correlation spectroscopy (FCCS) for the detection and discrimination between nanobeads labeled with different fluorescent dyes [69]. The chip has been successfully employed to detect a single bacteriophage Qβ virus without the need for particle immobilization [70]. The achieved sensitivity is approaching detection of virus on single level and is comparable to sensitivity reported with other microfluidic chips [71–73] and other integrated techniques, such as evanescent waveguide sensors [74–76], electrical nanowire arrays [77], and nanoelectromechanical cantilevers [78]. More recently, a solid-state nanopore has been integrated in the optofluidic chip. The nanopore ensures controlled delivery of individual bioparticles into the ARROW optofluidic channels. The bioparticle detection involved both electrical and optical measurements. Using this arrangement, the authors demonstrated single-particle detection of a fluorescently-labeled influenza virus and λ-DNA [79,80]. Optical detection techniques that complement the electrical measurements are well-known as very powerful, high-sensitive tools for single-particle detection using nanopore sensing [81]. However, the proposed optofluidic chip presented the advantage of multi-level integration of fluidic, optical, and electrical control on a single chip.

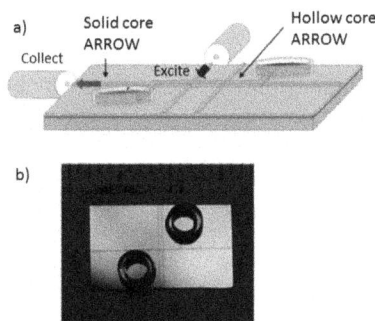

Figure 7. Planar optofluidic chip. (**a**) Schematic view of the integrated optofluidic chip showing hollow- and solid-core ARROW waveguides and (**b**) photograph of fabricated chip. Reprinted with permission from [68]. Copyright (2007) RSC Publishing.

Manipulation of bio-particles in a fluid sample is of great interest in the field of optofluidics as it enables analysis of biomolecules at a single level [82]. Basically, by using the chip design proposed in [68], very interesting operations have been demonstrated based on an all-optical particle control via the integrated waveguides, *i.e.*, the trapping and sorting of the particles in a fluid flow [83–86]. The planar nature of the optofluidic chip suggest that this technique could be applied for sample processing of a liquid with suspended bioparticles (viruses, bacteria, cells, and other microorganisms) for on-chip single particles for optical sensing and analysis. The authors exploited the combination of optical trapping and fluorescence detection of single microorganisms by evaluating the photobleaching dynamics of stained DNA in *Escherichia coli* bacteria [85].

In striving for miniaturization, several efforts have focused on the integration of various optical components on chip, such as slits and lenses [87,88], mirrors [89], filters [90,91], *etc.* Spectral filters are commonly used to reject pump contribution and distinguish the sample signal from background noise. Integration of optical filters on planar platforms can greatly support device miniaturization for high sensitivity in spectroscopic techniques, avoiding the use of bulky optical components [90]. In an ARROW waveguide, wavelength filtering can be achieved by exploiting the inherent spectral dependence of the confinement mechanism [11]. In the work of Philips *et al.* [92], an integrated notch filter has been implemented on a planar ARROW optofluidic chip for spectroscopic applications. The interference claddings of the collection solid waveguides have been suitably designed in order to confine the analyte signal wavelengths while rejecting the excitation light. In the work of Measor *et al.* [93] a similar approach has been used to tailor the spectral response of both liquid and solid core ARROWs. The authors demonstrated the ability to simultaneously reject the pump light while preserving low loss propagation of the analyte signal in Forster resonance energy transfer (FRET) detection of doubly-labeled oligonucleotides. High extinction of 37 dB was shown with a 4 mm long optofluidic filter. The estimated LOD of 15 nM is comparable with other integrated approaches [91].

In the examples presented so far, although the high level of optofluidic integration, the microfluidic functionality is essentially restricted to the ARROW ability to host fluids. On-chip devices for complete bio-medical and chemical analysis, however, often require more sophisticated microfluidic functions, *i.e.*, to accurately mix fluids and control their flow rates. Optofluidics arises from the need to move these functionalities on-chip without sacrificing the optical part. Actually, optofluidics attempts to find design solutions to suitably merge fluidic and photonic elements and improve the overall device functionalities. It is also becoming even more recognized that microfluidic systems can be more easily fabricated using polymer materials and related fabrication technologies. This choice benefits from the reduced cost and simpler fabrication procedures, which could promote larger-scale diffusion. On the basis of these concerns, ARROW-based optofluidic chips that make use of a hybrid technology to integrate microfluidic system have been recently developed by exploiting both surface and bulk micromachining fabrication procedures.

Based on bulk micromachining process, a hybrid ARROW optofluidic platform has been developed using the combination of PDMS and silicon-based materials. PDMS is extensively used in microfluidic technology because of its optical transparency in the visible/UV region, the biocompatibility and the low-cost fabrication procedures, like soft lithography [94]. The opportunity of combining silicon and polymer technologies has been firstly demonstrated with the fabrication of a single-mode h-ARROW, as reported in the work of Testa *et al.* [28]. Starting from this vision, the optofluidic platform has been assembled in a modular structure, with the upper polymer parts having both optical and microfluidic functions [95] (Figure 8). In particular, by moving from the bottom to the top of the structure, in the first PDMS layer (layer a) hybrid solid-core ARROWs have been integrated to deliver the optical signal towards the collection optical fibers. They also sealed the open-ended channels, preventing liquid evaporation. Merely fluid handling functionalities were restricted to the second PDMS layer (layer b), where a microfluidic mixer was incorporated. Such layer can be easily replaced with another, in order to adapt the microfluidic system to specific sensing applications. This confers to the platform high design flexibility. A multimode h-ARROW was used in order to allow self-aligned positioning of off-chip fibers with solid waveguides through the optofluidic channels. The sensing performance of the device was tested by performing fluorescence measurements at different dye concentrations, controlled by means of the integrated micromixer; an LOD of 2.5 nM was demonstrated. Even if this value is high as compared with the current state of the art of fluorescence measurement using evanescent waveguide sensors [96], the proposed modular approach for chip assembling into a compact and multifunctional format is very promising and offers prospects of an increased flexibility. For comparison, a modular approach for slot biochemical sensors have also been recently proposed in the work of Carlborg *et al.* [52] An array of optical slot-waveguide ring resonator sensors has been packaged into a chip incorporating microfluidic sample handling, with a microfluidic layer assembled in a compact cartridge. The authors demonstrated a volume refractive index detection limit of 5×10^{-6} RIU and surface mass detection limit of 0.9 pg/mm^2.

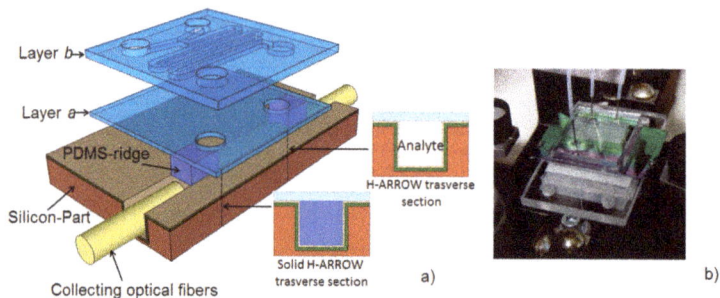

Figure 8. (**a**) Schematic of the proposed hybrid ARROW optofluidic platform with transverse section of liquid h-ARROW and solid h-ARROW; (**b**) Photograph of fabricated chip. Reprinted with permission from [95]. Copyright (2014) OSA.

A hybrid ARROW optofluidic platform has been also proposed in the work of Parks *et al.* [97]. In this configuration, the fluidic and optical chips were two separate entities, connected via metal reservoirs as shown in Figure 9a. The fluidic layer was fabricated in PDMS by soft lithography, being aligned but not in contact with the optical layer. This allows great design reconfigurability, as the PDMS layer can be removed and replaced without losing the alignment of the fluidic reservoirs with the optical chip. Microfluidic capabilities of fluidic mixing, distribution, and filtering have been demonstrated in combination with single-molecule fluorescent detection of dye-labeled λ-DNA. Using a similar approach for microfluidic integration, a more sophisticated fluid handling system capable of advanced microfluidic functions has been integrated with the optofluidic chip (Figure 9b). In particular, an actively programmable microfluidic layer made of PDMS was obtained by employing an interconnected array of nanoliter microvalves. Precise sample handling functions such as mixing, splitting, delivering, and storing were implemented for molecular diagnosis of viral nucleic acids from complex solutions [98], and for amplification-free detection and quantification of Ebola virus on clinical samples [99]. By using the advantages of the integrated PDMS-based microfluidic chip (automaton) to process the fluid sample, the authors demonstrated a LOD of 0.2 pfu/mL, comparable with other amplification-free methods like PCR analysis [100] and more than four orders of magnitude lower than other chip-based approaches [101].

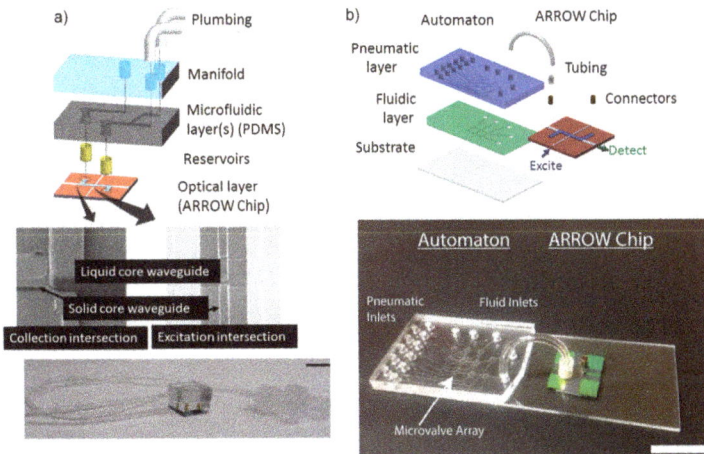

Figure 9. (**a**) Expanded view of PDMS integration with an ARROW optofluidic chip. Pictures are SEM images of optical intersections. Reprinted with permission from [97]. Copyright (2013) RSC Publishing. (**b**) Schematic of hybrid integration of automaton and ARROW chips. Reprinted with permission from [98]. Copyright (2014) AIP Publishing.

5. Conclusions

The discussion in this review demonstrates that the liquid core ARROW is a valuable tool for optofluidic integration. ARROWs have been successfully fabricated using current silicon technologies, showing very high optical performances. With core sections of a few microns, these waveguides can contain ultra-small fluid volumes of nanoliters to a few picoliters and simultaneously analyze it down to the single-molecule detection. As demonstrated by discussed examples, ARROWs have the capability to enhance the sensitivity of detection in several optical methods of analysis like fluorescence, Raman spectroscopy, and refractometry. Using ARROWs, complex interferometric sensing devices like a Mach–Zehnder interferometer or ring resonators have been designed and successfully tested.

Integrated optofluidic platforms have been demonstrated by combining solid core with liquid core ARROWs and microfluidic devices in several planar configurations. This represents a significant step forward from the previous version of an ARROW platform, opening a path towards an effective realization of chip-scale integrated system. The integration of planar optical components with microfluidic tools at the chip-scale represents the future trend of sensing systems for point-of care diagnosis or *in situ* monitoring. Increasing the synergy between these technologies can lead to highly reliable and portable systems.

By using ARROW optofluidic platforms, several advanced sensing methods have been successfully implemented for the manipulation, detection and analysis of single bioparticles directly on chip. The integration of microfluidic systems capable of advanced operations into an ARROW optofluidic chip has been the topic of intense investigation in recent years. Novel solutions have been proposed that are based on hybrid silicon/polymer technologies. The use of polymer to fabricate the microfluidic system has the potential to address the requirements of reduced fabrication cost and simplified manufacturing. The possibility to realize an optofluidic infrastructure of interconnected solid- and liquid-core waveguides on a planar silicon substrate and the hybrid integration of the ARROW platform has been demonstrated to be very attractive and creates a pathway for further integration of optoelectronic components towards a complete and self-contained chip integrated system. Based on the great improvements obtained in recent years, further progresses can be envisioned in fabrication methods that will improve the overall functionalities and promote commercial potential of ARROW-based optofluidic chip.

Acknowledgments: The research leading to these results has received the financial support of the Italian Minister of University and Research (MIUR), Futuro in Ricerca (FIR) programme under the grant N. RBFR122KL1 (SENS4BIO).

Conflicts of Interest: The authors declare no conflict of interest.

References

1. Monat, C.; Domachuk, P.; Eggleton, B.J. Integrated optofluidics: A new river of light. *Nat. Photon.* **2007**, *1*, 106–114. [CrossRef]
2. Psaltis, D.; Quake, S.R.; Yang, C. Developing optofluidic technology through the fusion of microfluidics and optics. *Nature* **2006**, *442*, 381–386. [CrossRef] [PubMed]
3. Fan, X.; White, I.M. Optofluidic microsystems for chemical and biological analysis. *Nat. Photon.* **2011**, *5*, 591–597. [CrossRef] [PubMed]
4. Testa, G.; Persichetti, G.; Bernini, R. Optofluidic Approaches for Enhanced Microsensor Performances. *Sensors* **2015**, *15*, 465–484. [CrossRef] [PubMed]
5. Schmidt, H.; Hawkins, A.R. Optofluidic waveguides: I. Concepts and implementations. *Microfluid. Nanofluid.* **2008**, *4*, 3–16. [CrossRef] [PubMed]
6. Bernini, R.; Campopiano, S.; Zeni, L. Silicon micromachined hollow optical waveguides for sensing applications. *IEEE J. Sel. Top. Quant. Electron.* **2002**, *8*, 106–110. [CrossRef]
7. Erickson, D.; Rockwood, T.; Emery, T.; Scherer, A.; Psaltis, D. Nanofluidic tuning of photonic crystal circuits. *Opt. Lett.* **2006**, *31*, 59–61. [CrossRef] [PubMed]
8. Surdo, S.; Merlo, S.; Carpignano, F.; Strambini, L.M.; Trono, C.; Giannetti, A.; Baldini, F.; Barillaro, G. Optofluidic microsystems with integrated vertical one-dimensional photonic crystals for chemical analysis. *Lab Chip* **2012**, *12*, 4403–4415. [CrossRef] [PubMed]
9. Almeida, V.R.; Xu, Q.; Barrios, C.A.; Lipson, M. Guiding and confining light in void nanostructure. *Opt. Lett.* **2004**, *29*, 1209–1211. [CrossRef] [PubMed]
10. Barrios, C.A. Optical Slot-Waveguide Based Biochemical Sensors. *Sensors* **2009**, *9*, 4751–4765. [CrossRef] [PubMed]
11. Schmidt, H.; Yin, D.; Barber, J.P.; Hawkins, A.R. Hollow-core waveguides and 2-D waveguide arrays for integrated optics of gases and liquids. *IEEE J. Sel. Top. Quant. Electron.* **2005**, *11*, 519–527. [CrossRef]
12. Delonge, T.; Fouckhardt, H. Integrated optical detection cell based on Bragg reflecting waveguides. *J. Chromatogr. A* **1995**, *716*, 135–139. [CrossRef]

13. Baba, T.; Kokubun, Y. Dispersion and radiation loss characteristics of antiresonant reflecting optical waveguides – numerical results and analytical expressions. *IEEE J. Quantum. Electron.* **1992**, *28*, 1689–1700. [CrossRef]

14. Litchinitser, N.M.; Abeeluck, A.K.; Headley, C.; Eggleton, B.J. Antiresonant reflecting photonic crystal optical waveguides. *Opt. Lett.* **2002**, *27*, 1592–1594. [CrossRef] [PubMed]

15. Archambault, J.-L.; Black, R.J.; Lacroix, S.; Bures, J. Loss calculations for antiresonant waveguides. *J. Lightwave Technol.* **1993**, *11*, 416–423. [CrossRef]

16. Testa, G.; Persichetti, G.; Bernini, R. Design and optimization of an optofluidic ring resonator based on liquid-core hybrid ARROWs. *IEEE Photon. J.* **2014**, *6*, 2201614. [CrossRef]

17. Bernini, R.; De Nuccio, E.; Minardo, A.; Zeni, L.; Sarro, P.M. Integrated silicon optical sensors based on hollow core waveguide. *Proc. SPIE* **2007**, *6477*, 647714.

18. Gollub, A.H.; Carvalho, D.O.; Paiva, T.C.; Alayo, M.I. Hollow core ARROW waveguides fabricated with SiO_xN_y films deposited at low temperatures. *Phys. Status Solidi C* **2010**, *7*, 964–967.

19. Yin, D.; Deamer, D.W.; Schmidt, H.; Barber, J.P.; Hawkins, A.R. Integrated optical waveguides with liquid cores. *Appl. Phys. Lett.* **2004**, *85*, 3477. [CrossRef]

20. Holmes, M.; Keeley, J.; Hurd, K.; Schmidt, H.; Hawkins, A.R. Optimized pirhana etching process for SU8-based MEMS and MOEMS construction. *J. Micromech. Microeng.* **2010**, *20*, 115008. [CrossRef] [PubMed]

21. Yin, D.; Schmidt, H.; Barber, J.P.; Hawkins, A.R. Integrated ARROW waveguides with hollow cores. *Opt. Express* **2004**, *12*, 2710–2715. [CrossRef] [PubMed]

22. Yin, D.; Barber, J.P.; Hawkins, A.R.; Schmidt, H. Waveguide loss optimization in hollow-core ARROW waveguides. *Opt. Express* **2005**, *13*, 9331–9336. [CrossRef] [PubMed]

23. Yin, D.; Barber, J.P.; Hawkins, A.R.; Schmidt, H. Highly efficient fluorescence detection in picoliter volume liquid-core waveguides. *Appl. Phys. Lett.* **2005**, *87*, 211111. [CrossRef]

24. Testa, G.; Huang, Y.; Zeni, L.; Sarro, P.M.; Bernini, R. Liquid core ARROW waveguides by atomic layer deposition. *IEEE Photon. Technol. Lett.* **2010**, *22*, 616–618. [CrossRef]

25. Campopiano, S.; Bernini, R.; Zeni, L.; Sarro, P.M. Microfluidic sensor based on integrated optical hollow waveguides. *Opt. Lett.* **2004**, *29*, 1894–1896. [CrossRef] [PubMed]

26. Barber, J.P.; Lunt, E.J.; George, Z.A.; Yin, D.; Schmidt, H.; Hawkins, A.R. Integrated hollow waveguide with arc shaped core. *IEEE Photon. Technol. Lett.* **2006**, *18*, 28–30. [CrossRef]

27. Yin, D.; Schmidt, H.; Barber, J.P.; Lunt, E.J.; Hawkins, A.R. Optical characterization of arch-shaped ARROW waveguides with liquid cores. *Opt. Express* **2005**, *13*, 10564–10570. [CrossRef] [PubMed]

28. Testa, G.; Huang, Y.; Zeni, L.; Sarro, P.M.; Bernini, R. Hybrid Silicon-PDMS optofluidic ARROW waveguide. *IEEE Photon. Technol. Lett.* **2012**, *24*, 1307–1309. [CrossRef]

29. Bernini, R.; De Nuccio, E.; Brescia, F.; Minardo, A.; Zeni, L.; Sarro, P.M.; Palumbo, R.; Scarfi, M.R. Development and characterization of an integrated silicon micro flow cytometer. *Anal. BioAnal. Chem.* **2006**, *386*, 1267–1272. [CrossRef] [PubMed]

30. Measor, P.; Seballos, L.; Yin, D.; Barber, J.P.; Hawkins, A.R.; Zhang, J.; Schmidt, H. Integrated ARROW waveguides for molecule specific surface-enhanced Raman sensing. In Proceedings of the 2006 16th Biennial University/Government/Industry Microelectronics Symposium, San Jose, CA, USA, 25–28 June 2006; pp. 181–182.

31. Bernini, R.; Testa, G.; Zeni, L.; Sarro, P.M. A 2 × 2 Optofluidic multimode interference coupler. *IEEE J. Sel. Top. Quantum Electron.* **2009**, *15*, 1478–1484. [CrossRef]

32. Testa, G.; Huang, Y.; Sarro, P.M.; Zeni, L.; Bernini, R. High-visibility optofluidic Mach–Zehnder interferometer. *Opt. Lett.* **2010**, *35*, 1584–1586. [CrossRef] [PubMed]

33. Testa, G.; Collini, C.; Lorenzelli, L.; Bernini, R. Planar silicon-polydimethylsiloxane optofluidic ring resonator sensors. *IEEE Photon. Tech. Lett.* **2016**, *28*, 155–158. [CrossRef]

34. Estevez, M.-C.; Alvarez, M.; Lechuga, L.M. Integrated optical devices for lab-on-a-chip biosensing applications. *Laser Photonics Rev.* **2012**, *6*, 463–487. [CrossRef]

35. Prieto, F.; Sepúlveda, B.; Calle, A.; Llobera, A.; Domínguez, C.; Abad, A.; Montoya, A.; Lechuga, L.M. An integrated optical interferometric nanodevice based on silicon technology for biosensor applications. *Nanotechnology* **2003**, *14*, 907–912. [CrossRef]

36. Nitkowski, A.; Baeumner, A.; Lipson, M. On-chip spectrophotometry for bioanalysis using microring resonators. *Biomed. Opt. Express* **2011**, *2*, 271–277. [CrossRef] [PubMed]

37. Chin, L.K.; Liu, A.Q.; Soh, Y.C.; Lim, C.S.; Lin, C.L. A reconfigurable optofluidic Michelson interferometer using tunable droplet grating. *Lab Chip* **2010**, *10*, 1072–1078. [CrossRef] [PubMed]
38. Grillet, C.; Domachuk, P.; Ta'eed, V.; Mägi, E.; Bolger, J.A.; Eggleton, B.J.; Rodd, L.E.; Cooper-White, J. Compact tunable microfluidic interferometer. *Opt. Express* **2004**, *12*, 5440–5447. [CrossRef] [PubMed]
39. Domachuk, P.; Grillet, C.; Ta'eed, V.; Mägi, E.; Bolger, J.; Eggleton, B.J.; Rodd, L.E.; Cooper-White, J. Microfluidic interferometer. *Appl. Phys. Lett.* **2005**, *86*, 024103. [CrossRef]
40. Ozcelik, D.; Parks, J.W.; Wall, T.A.; Stott, M.A.; Cai, H.; Parks, J.W.; Hawkins, A.R.; Schmidt, H. Optofluidic wavelength division multiplexing for single-virus detection. *Proc. Natl. Acad. Sci. USA* **2015**, *112*, 12933–12937. [CrossRef] [PubMed]
41. Testa, G.; Huang, Y.; Sarro, P.M.; Zeni, L.; Bernini, R. Integrated optofluidic ring resonator. *Appl. Phys. Lett.* **2010**, *97*, 131110. [CrossRef]
42. Bernini, R.; Testa, G.; Zeni, L.; Sarro, P.M. Integrated optofluidic Mach–Zehnder interferometer based on liquid core waveguides. *Appl. Phys. Lett.* **2008**, *93*, 011106. [CrossRef]
43. Dumais, P.; Callender, C.L.; Noad, J.P.; Ledderhof, C.J. Integrated optical sensor using a liquid-core waveguide in a Mach–Zehnder interferometer. *Opt. Express* **2008**, *16*, 18164–18172. [CrossRef] [PubMed]
44. Crespi, A.; Gu, Y.; Ngamsom, B.; Hoekstra, H.J.W.M.; Dongre, C.; Pollnau, M.; Ramponi, R.; van den Vlekkert, H.H.; Watts, P.; Cerullo, G.; *et al.* Three-dimensional Mach–Zehnder interferometer in a microfluidic chip for spatially-resolved label-free detection. *Lab Chip* **2010**, *10*, 1167–1173. [CrossRef] [PubMed]
45. Lapsley, M.I.; Chiang, I.-K.; Zheng, Y.B.; Ding, X.; Mao, X.; Huang, T.J. A single-layer, planar, optofluidic Mach–Zehnder interferometer for label-free detection. *Lab Chip* **2011**, *11*, 1795–1800. [CrossRef] [PubMed]
46. Kozma, P.; Kehl, F.; Ehrentreich-Förster, E.; Stamm, C.; Bier, F.F. Integrated planar optical waveguide interferometer biosensors: A comparative review. *Biosens. Bioelectron.* **2014**, *58*, 287–307. [CrossRef] [PubMed]
47. Mitra, A.; Deutsch, B.; Ignatovich, F.; Dykes, C.; Novotny, L. Nano-optofluidic Detection of Single Viruses and Nanoparticles. *ACS Nano* **2010**, *4*, 1305–1312. [CrossRef] [PubMed]
48. Huang, G.; Bolaños Quiñones, V.A.; Ding, F.; Kiravittaya, S.; Mei, Y.; Schmidt, O.G. Rolled-up optical microcavities with subwavelength wall thicknesses for enhanced liquid sensing applications. *ACS Nano* **2010**, *4*, 3123–3130. [CrossRef] [PubMed]
49. Berneschi, S.; Farnesi, D.; Cosi, F.; Conti, G.N.; Pelli, S.; Righini, G.C.; Soria, S. High Q silica microbubble resonators fabricated by arc discharge. *Opt. Lett.* **2011**, *36*, 3521–3523. [CrossRef] [PubMed]
50. White, I.M.; Oveys, H.; Fan, X. Liquid-core optical ring-resonator sensors. *Opt. Lett.* **2006**, *31*, 1319–1321. [CrossRef] [PubMed]
51. Li, H.; Fan, X. Characterization of sensing capability of optofluidic ring resonator biosensors. *Appl. Phys. Lett.* **2010**, *97*, 011105. [CrossRef]
52. Carlborg, C.F.; Gylfason, K.B.; Kaźmierczak, A.; Dortu, F.; Bañuls Polo, M.J.; Maquieira Catala, A.; Kresbach, G.M.; Sohlström, H.; Moh, T.; Vivien, L.; *et al.* A packaged optical slot-waveguide ring resonator sensor array for multiplex label-free assays in labs-on-chips. *Lab Chip* **2010**, *10*, 281–290. [CrossRef] [PubMed]
53. Zhu, H.; Dale, P.S.; Caldwell, C.W.; Fan, X. Rapid and Label-Free Detection of Breast Cancer Biomarker CA15-3 in Clinical Human Serum Samples with Optofluidic Ring Resonator Sensors. *Anal. Chem.* **2009**, *81*, 9858–9865. [CrossRef] [PubMed]
54. Fei, P.; Chen, Z.; Men, Y.; Li, A.; Shen, Y.; Huang, Y. A compact optofluidic cytometer with integrated liquid-core/PDMS-cladding waveguides. *Lab Chip* **2012**, *12*, 3700–3706. [CrossRef] [PubMed]
55. Fleger, M.; Neyer, A. PDMS microfluidic chip with integrated waveguides for optical detection. *Microelectron. Eng.* **2006**, *83*, 1291–1293. [CrossRef]
56. Mogensen, K.B.; El-Ali, J.; Wolff, A.; Kutter, J.P. Integration of polymer waveguides for optical detection in microfabricated chemical analysis systems. *Appl. Opt.* **2003**, *42*, 4072–4079. [CrossRef] [PubMed]
57. Friis, P.; Hoppe, K.; Leistiko, O.; Mogensen, K.B.; Hübner, J.; Kutter, J.P. Monolithic integration of microfluidic channels and optical waveguides in silica on silicon. *Appl. Opt.* **2001**, *40*, 6246–6251. [CrossRef] [PubMed]
58. Wang, Z.; Yan, H.; Wang, Z.; Zou, Y.; Yang, C.-J.; Chakravarty, S.; Subbaraman, H.; Tang, N.; Xu, X.; Fan, D.L.; *et al.* Ultralow-loss waveguide crossings for the integration of microfluidics and optical waveguide sensors. *Proc. SPIE* **2015**, *9320*, 932012.

59. Yin, D.; Deamer, D.W.; Schmidt, H.; Barber, J.P.; Hawkins, A.R. Single-molecule detection sensitivity using planar integrated optics on a chip. *Opt. Lett.* **2006**, *31*, 2136–2138. [CrossRef] [PubMed]

60. Measor, P.; Seballos, L.; Yin, D.; Zhang, J.Z.; Lunt, E.J.; Hawkins, A.R.; Schmidt, H. On-chip surface-enhanced Raman scattering detection using integrated liquid-core waveguides. *Appl. Phys. Lett.* **2007**, *90*, 211107. [CrossRef]

61. Li, Q.-L.; Li, B.-W.; Wang, Y.-Q. Surface-enhanced Raman scattering microfluidic sensor. *RSC Adv.* **2013**, *3*, 13015–13026. [CrossRef]

62. Wang, M.; Jing, N.; Chou, I.-H.; Cote, G.L.; Kameoka, J. An optofluidic device for surface enhanced Raman spectroscopy. *Lab Chip* **2007**, *7*, 630–632. [CrossRef] [PubMed]

63. Lunt, E.J.; Phillips, B.S.; Jones, C.J.; Hawkins, A.R.; Measor, P.; Kuehn, S.; Schmidt, H. Hollow waveguide optimization for fluorescence based detection. *Proc. SPIE* **2008**, *6883*, 68830H.

64. Lunt, E.J.; Wu, B.; Keeley, J.M.; Measor, P.; Schmidt, H.; Hawkins, A.R. Hollow ARROW Waveguides on Self-Aligned Pedestals for Improved Geometry and Transmission. *IEEE Photonics Technol. Lett.* **2010**, *22*, 1147–1149. [CrossRef] [PubMed]

65. Zhao, Y.; Jenkins, M.H.; Measor, P.; Leake, K.; Liu, S.; Schmidt, H.; Hawkins, A.R. Hollow waveguides with low intrinsic photoluminescence fabricated with Ta_2O_5 and SiO_2 film. *Appl. Phys. Lett.* **2011**, *98*, 091104. [CrossRef] [PubMed]

66. Phillips, B.S.; Jenkins, M.H.; Liu, S.; Schmidt, H.; Hawkins, A.R. Selective Thin-Film deposition for optofluidic platforms with optimized transmission. *IEEE Photon. Technol. Lett.* **2011**, *23*, 721–723. [CrossRef]

67. Zhao, Y.; Leake, K.D.; Measor, P.; Jenkins, M.H.; Keeley, J.; Schmidt, H.; Hawkins, A.R. Optimization of interface transmission between integrated solid core and optofluidic waveguides. *IEEE Photon. Technol. Lett.* **2012**, *24*, 46–48. [CrossRef]

68. Yin, D.; Lunt, E.J.; Rudenko, M.I.; Deamer, D.W.; Hawkins, A.R.; Schmidt, H. Planar optofluidic chip for single particle detection, manipulation, and analysis. *Lab Chip* **2007**, *7*, 1171–1175. [CrossRef] [PubMed]

69. Chen, A.; Eberle, M.M.; Lunt, E.J.; Liu, S.; Leake, K.; Rudenko, M.I.; Hawkins, A.R.; Schmidt, H. Dual-color fluorescence cross-correlation spectroscopy on a planar optofluidic chip. *Lab Chip* **2011**, *11*, 1502–1506. [CrossRef] [PubMed]

70. Rudenko, M.I.; Kühn, S.; Lunt, E.J.; Deamer, D.W.; Hawkins, A.R.; Schmidt, H. Ultrasensitive Qβ phage analysis using fluorescence correlation spectroscopy on an optofluidic chip. *Biosens. Bioelectron.* **2009**, *24*, 3258–3263. [CrossRef] [PubMed]

71. Bao, N.; Wang, J.; Lu, C. Recent advances in electric analysis of cells in microfluidic systems. *Anal. Bioanal. Chem.* **2008**, *391*, 933–942. [CrossRef] [PubMed]

72. Choi, J.; Kang, M.; Jung, J.H. Integrated micro-optofluidic platform for real-time detection of airborne microorganisms. *Sci. Rep.* **2015**, *5*, 15983. [CrossRef] [PubMed]

73. Liu, Y.S.; Walter, T.M.; Chang, W.J.; Lim, K.S.; Yang, L.; Lee, S.W.; Aronson, A.; Bashir, R. Electrical detection of germination of viable model *Bacillus anthracis* spores in microfluidic biochips. *Lab Chip* **2007**, *7*, 603–610.

74. Mukundan, H.; Anderson, A.S.; Grace, W.K.; Grace, K.M.; Hartman, N.; Martinez, J.S.; Swanson, B.I. Waveguide-Based Biosensors for Pathogen Detection. *Sensors* **2009**, *9*, 5783–5809. [CrossRef] [PubMed]

75. Martinez, J.S.; Grace, W.K.; Grace, K.M.; Hartman, N.; Swanson, B.I. Pathogen detection using single mode planar optical waveguides. *J. Mater. Chem.* **2005**, *15*, 4639–4647. [CrossRef]

76. Donaldson, K.A.; Kramer, M.F.; Lim, D.V. A rapid detection method for Vaccinia virus, the surrogate for smallpox virus. *Biosens. Bioelectron.* **2004**, *20*, 322–327. [CrossRef] [PubMed]

77. Patolsky, F.; Zheng, G.; Hayden, O.; Lakadamyali, M.; Zhuang, X.; Lieber, C.M. Electrical detection of single viruses. *Proc. Natl. Acad. Sci. USA* **2004**, *101*, 14017–14022. [CrossRef] [PubMed]

78. Ilic, B.; Yang, Y.; Craighead, H.G. Virus detection using nanoelectromechanical devices. *Appl. Phys. Lett.* **2004**, *85*, 2604. [CrossRef]

79. Shuo, L.; Zhao, Y.; Parks, J.W.; Deamer, D.W.; Hawkins, A.R.; Schmidt, H. Correlated electrical and optical analysis of single nanoparticles and biomolecules on a nanopore-gated optofluidic chip. *Nano Lett.* **2014**, *14*, 4816–4820.

80. Shuo, L.; Wall, T.A.; Ozcelik, D.; Parks, J.W.; Hawkins, A.R.; Schmidt, H. Electro-optical detection of single λ-DNA. *Chem. Comm.* **2015**, *51*, 2084–2087.

81. Gilboa, T.; Meller, A. Optical sensing and analyte manipulation in solid-state nanopores. *Analyst* **2015**, *140*, 4733–4747. [CrossRef] [PubMed]

82. Kayani, A.A.; Khoshmanesh, K.; Ward, S.A.; Mitchell, A.; Kalantar-zadeh, K. Optofluidics incorporating actively controlled micro- and nano-particles. *Biomicrofluidics* **2012**, *6*, 031501. [CrossRef] [PubMed]
83. Kühn, S.; Measor, P.; Lunt, E.J.; Phillips, B.S.; Deamer, D.W.; Hawkins, A.R.; Schmidt, H. Loss-based optical trap for on-chip particle analysis. *Lab Chip* **2009**, *9*, 2212–2216. [CrossRef] [PubMed]
84. Kühn, S.; Lunt, E.J.; Phillips, B.S.; Hawkins, A.R.; Schmidt, H. Optofluidic particle concentration by a long-range dual-beam trap. *Opt. Lett.* **2009**, *34*, 2306–2308. [CrossRef] [PubMed]
85. Kühn, S.; Phillips, B.S.; Lunt, E.J.; Hawkins, A.R.; Schmidt, H. Ultralow power trapping and fluorescence detection of single particles on an optofluidic chip. *Lab Chip* **2010**, *10*, 189–194. [CrossRef] [PubMed]
86. Leake, K.D.; Phillips, B.S.; Yuzvinsky, T.D.; Hawkins, A.R.; Schmidt, H. Optical particle sorting on an optofluidic chip. *Opt. Express* **2013**, *21*, 32605–32610. [CrossRef] [PubMed]
87. Schuhladen, S.; Banerjee, K.; Stürmer, M.; Müller, P.; Wallrabe, U.; Zappe, H. Variable optofluidic slit aperture. *Light Sci. Appl.* **2016**, *5*, e16005. [CrossRef]
88. Seow, Y.C.; Lim, S.P.; Lee, H.P. Optofluidic variable-focus lenses for light manipulation. *Lab Chip* **2012**, *12*, 3810–3815. [CrossRef] [PubMed]
89. Müller, P.; Kopp, D.; Llobera, A.; Zappe, H. Optofluidic router based on tunable liquid-liquid mirrors. *Lab Chip* **2014**, *14*, 737–743. [CrossRef] [PubMed]
90. Llobera, A.; Demming, S.; Joensson, H.N.; Vila-Planas, J.; Andersson-Svahn, H.; Büttgenbach, S. Monolithic PDMS passband filters for fluorescence detection. *Lab Chip.* **2010**, *10*, 1987–1992. [CrossRef] [PubMed]
91. Dandin, M.; Abshire, P.; Smela, E. Optical filtering technologies for integrated fluorescence sensors. *Lab Chip* **2007**, *7*, 955–977. [CrossRef] [PubMed]
92. Phillips, B.S.; Measor, P.; Zhao, Y.; Schmidt, H.; Hawkins, A.R. Optofluidic notch filter integration by lift-off of thin films. *Opt. Express* **2010**, *18*, 4790–4795. [CrossRef] [PubMed]
93. Measor, P.; Phillips, B.S.; Chen, A.; Hawkins, A.R.; Schmidt, H. Tailorable integrated optofluidic filters for biomolecular detection. *Lab Chip* **2011**, *11*, 899–904. [CrossRef] [PubMed]
94. Ng, J.M.; Gitlin, I.; Stroock, A.D.; Whitesides, G.M. Components for integrated poly(dimethylsiloxane) microfluidic systems. *Electrophoresis* **2002**, *23*, 3461–3473. [CrossRef]
95. Testa, G.; Persichetti, G.; Sarro, P.M.; Bernini, R. A hybrid silicon-PDMS optofluidic platform for sensing applications. *Biomed. Opt. Express* **2014**, *5*, 417–426. [CrossRef] [PubMed]
96. Taitt, C.R.; Anderson, G.P.; Ligler, F.S. Evanescent wave fluorescence biosensors. *Biosens. Bioelectron.* **2005**, *20*, 2470–2487. [CrossRef] [PubMed]
97. Parks, J.W.; Cai, H.; Zempoaltecatl, L.; Yuzvinsky, T.D.; Leake, K.; Hawkins, A.R.; Schmidt, H. Hybrid optofluidic integration. *Lab Chip* **2013**, *13*, 4118–4123. [CrossRef] [PubMed]
98. Parks, J.W.; Olson, M.A.; Kim, J.; Ozcelik, D.; Cai, H.; Carrion, R.; Patterson, J.L.; Mathies, R.A.; Hawkins, A.R.; Schmidt, H. Integration of programmable microfluidics and on-chip fluorescence detection for biosensing applications. *Biomicrofluidics* **2014**, *8*, 054111. [CrossRef] [PubMed]
99. Cai, H.; Parks, J.W.; Wall, T.A.; Stott, M.A.; Stambaugh, A.; Alfson, K.; Griffiths, A.; Mathies, R.A.; Carrion, R.; Patterson, J.L.; *et al.* Optofluidic analysis system for amplification-free, direct detection of Ebola infection. *Sci. Rep.* **2015**, *5*, 14494. [CrossRef] [PubMed]
100. Trombley, A.R.; Wachter, L.; Garrison, J.; Buckley-Beason, V.-A.; Jahrling, J.; Hensley, L.E.; Schoepp, R.J.; Norwood, D.A.; Goba, A.; Fair, J.N.; *et al.* Comprehensive panel of real-time TaqMan™ polymerase chain reaction assays for detection and absolute quantification of Filoviruses, Arenaviruses, and New World Hantaviruses. *Am. J. Trop. Med. Hyg.* **2010**, *82*, 954–960. [CrossRef] [PubMed]
101. Daaboul, G.G.; Lopez, C.A.; Chinnala, J.; Goldberg, B.B.; Connor, J.H.; Ünlü, M.S. Digital sensing and sizing of vesicular stomatitis virus pseudotypes in complex media: A model for Ebola and Marburg detection. *ACS Nano* **2014**, *8*, 6047–6055. [CrossRef] [PubMed]

micromachines

MDPI

Review

Opto-Microfluidic Immunosensors:
From Colorimetric to Plasmonic

Jie-Long He [†], Da-Shin Wang [†] and Shih-Kang Fan *

Department of Mechanical Engineering, National Taiwan University, Taipei 10617, Taiwan;
d93b47202@ntu.edu.tw (J.-L.H.); amydsw@gmail.com (D.-S.W.)
*   Correspondence: skfan@fan-tasy.org; Tel.: +886-2-3366-4946; Fax: +886-2-2363-1755
†   These authors contributed equally to this work.

Academic Editors: Ian Papautsky and Nam-Trung Nguyen
Received: 13 December 2015; Accepted: 4 February 2016; Published: 15 February 2016

Abstract: Optical detection has long been the most popular technique in immunosensing. Recent developments in the synthesis of luminescent probes and the fabrication of novel nanostructures enable more sensitive and efficient optical detection, which can be miniaturized and integrated with microfluidics to realize compact lab-on-a-chip immunosensors. These immunosensors are portable, economical and automated, but their sensitivity is not compromised. This review focuses on the incorporation and implementation of optical detection and microfluidics in immunosensors; it introduces the working principles of each optical detection technique and how it can be exploited in immunosensing. The recent progress in various opto-microfluidic immunosensor designs is described. Instead of being comprehensive to include all opto-microfluidic platforms, the report centers on the designs that are promising for point-of-care immunosensing diagnostics, in which ease of use, stability and cost-effective fabrication are emphasized.

Keywords: optical detection; microfluidics; immunosensing

---

1. Introduction

An immunosensor is an affinity-based sensor that exploits a specific interaction between an antigen and an antibody to detect and to quantify an analyte, which is typically an antigen that binds to an antibody immobilized on a sensor surface. In response to foreign molecules called antigens, the immune system produces antibodies to recognize antigens. Antigens are generally proteins, polysaccharides, small molecules (haptens) and even short peptides from an antigenic epitope, which might include parts of bacteria, viruses and other microorganisms. Monoclonal antibody (mAb) technology is an important scientific achievement; this process can produce many specific antibodies for immunoassays, which have been widely used for clinical diagnostics and basic biological research [1–5]. It is also a powerful tool for epidemiological surveillance of microorganisms, especially for influenza [6–9]. The binding strength between an antibody and its antigen can be represented with K_d, a dissociation parameter, which typically lies in the nanomolar (nM) range [10]. A strong and specific interaction is the basis of a traditional immunoassay [11] and an emerging microfluidic immunosensor [12–15]. The production of high-quality antibodies or the design of alternative binding molecules and structures [16,17] hence plays a central role in deciding the specificity and sensitivity of an immunosensor. A subsequent immobilization on a sensor surface also affects the effective affinity. Because of the myriad design schemes and the chemical nature of this issue [18–20], an optimization of binding affinity is beyond the scope of this review, but one should note that a subsequent choice of transducer, which converts the bio-recognition event to a measurable signal, provides an additional rather than a decisive effect on the overall sensitivity of an immunosensor.

In contrast to a conventional immunoassay, which requires several repetitive steps with reagents, an immunosensor integrated with microfluidics has the advantage of automating and simplifying the steps to speed the measurement. Automation, speed, sensitivity and stability are the general indicators for the improvement of a sensor performance. Among detection methods of all types in immunosensing, which include electrochemical [21–23], optical [24,25], microgravimetric [26] and thermometric detection [27], optical detection is the mostly popular technique. The reason is its large ratio of signal to noise, non-destructive operation and rapid reading.

This review focuses on the introduction of several optical detection techniques and the microfluidic designs to facilitate the optical detection on a sensor. In Sections 2 and 3 the working principles of labeled and label-free optical detection methods are reviewed; varied implementation with microfluidics is briefly introduced to help the reader understand how microfluidics can be integrated. Section 4 provides a detailed review of recent progress in microfluidic optical immunosensors. Despite numerous optical schemes that have been proposed for immunosensing in the past decade, few are commercialized and adopted in a clinical laboratory. An overview of recent advances in opto-microfluidic immunosensing have presented in Table 1. This review is hence not intended to be comprehensive to cover all recent progress in this field, but instead centers on the techniques widely adopted in medical laboratories and those emerging as potential alternatives in the future. The desired features for future medical application include instrumentation simplicity, less cost per assay, high sensitivity and specificity, less requisite sample volume and purity for assay, ease of multiplexingand, in particular, nonspecific background. Optical detections with signals produced by non-optical excitation such as enzyme-linked immunosorbent assay (ELISA) and electrochemiluminescence generally equip with simpler instrumentation while attaining high sensitivity; however, the multiplexing capability is compromised as not much signal modality can be separated for multiple analytes. Photoluminescence and surface plasmon resonance detection are particularly advantageous for multiplex screening, which require more sophisticated optical excitation and detection systems. To help the reader quickly get a flavor of these designs in advance, the major methods and their limits of detection are listed in Table 1.

Table 1. Comparison of various optical microfluidic immunosensing methods.

Detection Methods	Signal Generation	Label	Time Required	Microfluidic Device Types	LOD [a] Level	Ref.
Colormetric detection	Visible light illumination	Enzymatic catalyzed probes for ELISA	Minutes to hours	Microchannel	pg/mL	[28]
				Microchannel	pg/mL	[29]
				Strip-based	ng/mL	[30]
				Disc-based	ng/mL	[31]
Chemiluminescence	Chemical reaction	Enzymatic catalyzed probes for CL	Minutes	Opto-microfluidic	pg/mL	[32]
				Microchannel	pg/mL	[33]
				μPAD [b]	ng/mL	[34]
				Disc-based	ng/mL	[35]
				EWOD [c]	μg/mL	[36]
Electrochemiluminescence	Electrochemical reaction	ECL probes	Minutes	μPAD	ng/mL	[37]
				Microchannel	ng/mL	[38]
Photoluminescence	Optical excitation	Photoluminescent dyes	Real time	Microchannel	pg/mL	[39]
				Microchannel	pg/mL	[40]
				Microchannel	pg/mm^2	[41]
				Microchannel	ng/mL	[42]
				EWOD	μg/mL	[43]
Surface plasmon resonance	Optical excitation with evanescent waves	Label-free	Real time	Microchannel	ng/mL	[44]
				EWOD	ng/mL	[45,46]

Notes: [a] LOD: Limit of detection. [b] μPADs: Microfluidic paper-based analytical devices. [c] EWOD: Electrowetting-on-dielectric.

2. Immunosensing with Optical Probes

Detection with an optical probe is the most common method in immunosensing [47]. The binding of an analyte with an immobilized antibody can be detected through an attached optical probe.

An optical signal such as colorimetric detection, chemiluminescence (CL), electrochemiluminescence (ECL) and photoluminescence can be measured to detect analytes; this information is the physical basis for optical immunosensing. The optical signals are generated enzymatically or electrochemically or with optical excitation, depending on the choice of probe (See Figure 1).

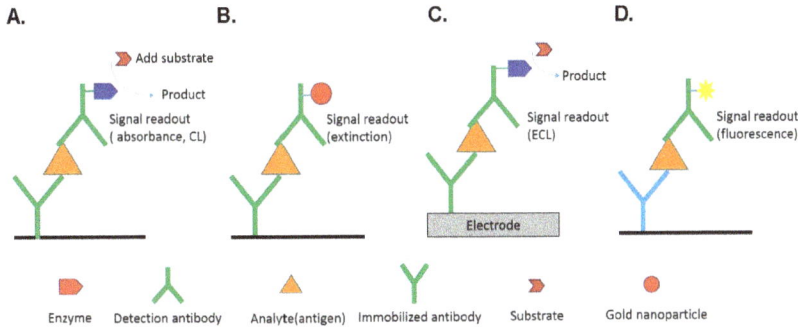

Figure 1. Comparison of immunosensing with optical probes. (**A**) Colorimetric and chemiluminescence (CL) induced by an enzyme; (**B**) Colorimetric changes by light absorption and scattering of a nanoparticle; (**C**) Electrochemiluminescence (ECL) from an electrochemically excited probe; (**D**) Photoluminescence from an optically excited probe.

2.1. Colorimetric Detection

The enzyme-linked immunosorbent assay (ELISA) that first appeared in the 1970s [48] used an enzyme rather than a radioactive label as the reporter. If the analyte is an antigen, an antibody is immobilized on the plate through adsorption. The detection is attained with another antibody conjugated (Figure 1A) with an enzyme that catalyzes the hydrolysis of a chromogenic substrate to render a colorimetric change measurable with a spectrophotometer, or simply visually. Common enzymes used in ELISA are horseradish peroxidase (HRP), alkaline phosphatase (AP) and β-D-galactosidase [28,49]. Washing and rinsing are incorporated after each incubation step to remove excess molecules and to eliminate non-specific binding. Apart from direct detection, ELISA schemes use a second antibody, which is a species-specific antibody that serves to recognize the first antibody. The second antibody is a detection antibody conjugated to the enzyme, as opposed to the first antibody in the direct method. Even though various solid-phase immunosensors are devised, ELISA performed on polystyrene microtiter plates is still the current standard in a medical laboratory; it is a sensitive technique with modest cost. The sensitivity arises from a signal amplification with enzyme-conjugates that allows the detection of minute concentrations of analytes, resulting in an improved detection limit to an ng level for biomarker detection [29,30,50].

The development of microfluidic ELISA has the advantages of saving the volume and duration of incubation of costly reagents. Lee *et al.* [31] developed a fully automated lab-on-a-disc (LOD) to perform microbead-based ELISA beginning with whole blood. An optical detector is equipped to quantify the colorimetric changes through absorbance. The use of microbeads increases the surface area and thus enhances the reaction efficiency, but microbeads are not easily retained in a suspension state; their transport hence cannot be accurately controlled with a continuous flow through a microfabricated channel. A disc platform provides an easy solution to this problem, in which the liquid transfer is controlled by the spin speed. With this LOD platform, the entire ELISA process from plasma blood separation, incubation of antibody, several steps of washing and final detection was attained automatically within 30 min using a drop of whole blood (150 μL), half the amount of a conventional ELISA. LOD immunosensing is hence an effective design for point-of-care (POC) laboratory settings [51].

2.2. Chemiluminescence (CL)

Chemiluminescence (CL) signifies light emitted in a chemical reaction of a CL molecule [52,53], as opposed to optical excitation as in the case of photoluminescence, in which the emission is induced with absorbed electromagnetic radiation (Figure 2A). The electron of an organic molecule is excited to a higher energy state, with the energy provided by the oxidation reaction rather than absorption of radiation, and then processes following excitation hold both for chemiluminescence and fluorescence or phosphorescence, together known as photoluminescence. This chemical reaction is initiated with enzyme catalysis like ELISA. As CL requires no optical excitation, problems associated with photoluminescence such as scattering of excitation light or source instability are absent, as is also non-selective excitation such as background autophotoluminescence. The detection simply involves collecting sufficient photons from an inexpensive photomultiplier tube; no additional cost for excitation lamp and associated optics is required with a CL-based immunosensor. Another advantage of chemiluminescence is the large linear dynamic range. As it is an emission process, the optical signal is linearly proportional to the concentration; the linear dynamic range extends from a minimum detectable concentration up to six orders of magnitude [52]. For monoplex assay, CL detection with a photomultiplier tube is hence a superior choice for optical immunosensing in terms of sensitivity and cost [54–56].

A common chemiluminescent probe is luminol or acridinium ester [57]. The sensitivity might be increased through an accumulation of luminescent products and amplification of the enzymatic transduction signal [58,59]. The chemiluminescent reaction requires peroxide to attain an intermediate state; this intermediate decomposes with simultaneous light emission (Figure 2B). The quantum efficiency of this reaction is small, about 1%. Enzymatic catalysis is a way to accelerate the reaction and it thereby increases the rate of light emission. The enzymatic catalysis notably does not affect the quantum efficiency of the reaction, but simply increases the rate of the reaction cycle so that more light is accumulated within a given interval. Typical enzymes to catalyze the reaction are microperoxidase (MPO), myeloperoxidase and horseradish peroxidase (HRP). The CL immunosensing scheme is analogous to that of ELISA with the detection antibody conjugated to an enzyme. The chemiluminescent probe serves as the substrate to be catalyzed, as shown in Figure 1A.

Chemiluminescence is a promising detection method for immunosensing because of its simplicity, great sensitivity, small cost [34,60] and low-power demands [1,61]. The challenge for it to be implemented in a microfluidic device is the weak light from a small volume, which requires a sensitive optical detector such as a photomultiplier tube and a finely adjusted optical arrangement to acquire the signals. Pires *et al.* [32] reported a simple inexpensive opto-microfluidic device modified with gold nanoparticles that was equipped with a sensitive and stable organic photodetector. The gold nanoparticles enhanced the CL of the HRP-Luminol complex to attain a sensitivity 2.5 pg/mL, which is 200 times as sensitive as current CL immunosensors. Signal enhancement, more sensitive detectors and improved optical alignment and focusing are important factors to be developed for the incorporation of CL detection in a microfluidic device.

Yao *et al.* [33] integrated a magnetic microparticle technique and a CL immunoassay to develop an automated microfluidic system for rapid detection of insulin concentration. This system is expected to become a revolutionary platform in the clinical diagnosis of diabetes. Guo *et al.* [35] integrated a CD-type microfluidic platform and a CL immunoassay to detect a kind of endocrine disruptor, alkylphenolpolyethoxylates (APnEO). Zeng *et al.* [62] reported a portable prototype of a CL detector on an electrowetting-on-dielectric (EWOD) digital microfluidic platform, which has potential for development as being a highly sensitive, cheap and portable immunodetector.

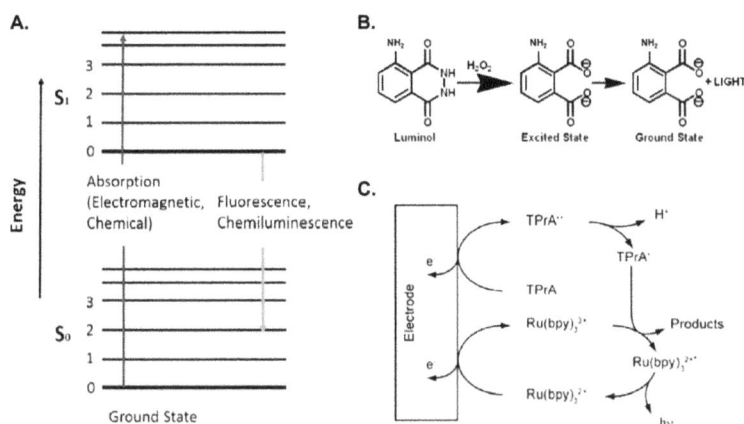

Figure 2. Mechanisms of luminescence. (**A**) Absorption of electromagnetic radiation, photoluminescence and chemiluminescence; (**B**) chemiluminescence; (**C**) electrochemiluminescence. Figure 2B is reproduced with permission from the Chemical Connection website [63].

2.3. Electrochemiluminescence (ECL)

Electrochemiluminescence (ECL) [64–66] pertains to chemiluminescence initiated in an electrochemical reaction. Among inorganic compounds known to be electrochemiluminescent, a widely exploited compound is a ruthenium complex, $Ru(bpy)_3^{2+}$. $Ru(bpy)_3^{2+}$ first proceeds through a redox reaction to produce $Ru(bpy)_3^{3+}$ and $Ru(bpy)_3^{+}$. In the presence of a co-reactant such as Tripropylamine (TPrA), $Ru(bpy)_3^{3+}$ converts to $Ru(bpy)_3^{+}$ and yields $Ru(bpy)_3^{2+}$ in an excited state that relaxes through photon emission as shown in Figure 2C. The reaction is efficient and can occur in an aqueous buffered solution with impurities and near 23 °C. These properties make it a popular ECL probe for immunosensing.

ECL has several advantages over conventional CL measurements. Because the reactants for ECL are produced by the applied potential, switching the applied potential can control the initiation and the course of a CL reaction. The "on and off" of CL light allows a user to discern the background stray light and include it into a correction. As the reactants are generated near the electrode and allowed to react immediately, the emission is concentrated near the electrode surface, which provides an accurate position for the subsequent optical alignment and focusing for maximum sensitivity. The intermediates of ruthenium complex can be converted back to $Ru(bpy)_3^{2+}$ after emitting light and are readily recycled at the electrode. The regenerated $Ru(bpy)_3^{2+}$ continuously participates the CL reaction in an excess of Tripropylamine, thus producing sufficient photons for detection. In addition, the recycling of reagent decreases the amount of reagents to be consumed while producing more photons per measurement cycle. ECL detection is more sensitive with more controllability than the conventional CL.

Because of the popularity of ECL immunosensing in a medical laboratory, a portable microfluidic ECL immunosensor is a promising device for POC examination. Wang *et al.* [37] introduced a battery-triggered microfluidic paper-based multiplex electrochemiluminescent immunosensor. Two ECL labels, $Ru(bpy)_3^{2+}$ and carbon nanodots, were used; four markers were detected using only two screen-printed working electrodes. Another modality of ECL detection [38], ECL resonance energy transfer (ECL-RET), in which the energy of ECL donor ($Ru(bpy)_3^{2+}$) transfers to that of an ECL acceptor (CdS nanorod), was adopted and proved to be highly sensitive for immunosensing. When a $Ru(bpy)_3^{2+}$/antibody was captured with an antigen on the CdS nanorod, the donor's ECL emission decreased and a new emission signal emerged. Using a CdS nanorod as a spot site to capture the analyte and to initiate ECL-RET created a 64-site immunosensing array, making possible a multiplexed and sensitive ECL immunosensor.

2.4. Photoluminescence

The light emitted in photoluminescence is similar to chemiluminescence, which involves the transfer of electronic energy from the first excited state S_1 to the ground state S_0 with a photon emission in the visible light region (see Figure 2A). The difference lies in that photoluminescence is generated from an optical excitation. Through the nature of optical excitation, several problems must be solved to attain great sensitivity, which includes the separation of excitation and detection, the filtering of scattered light, background photoluminescence from optical components, cuvettes and reagents, and photoluminescent quenching [67]. As the quantum efficiency of photoluminescent emission is generally sufficiently large to produce enough photons for detection, it can serve directly as a probe without addition of an enzyme as in the case of CL [42,68]; this effect makes it suitable for imaging use and multiplex sensing [69–71]. There is also a large selection of photoluminescent probes, such as membrane-bound, calcium-sensitive, potentiometric photoluminescent dyes *etc.*, for varied uses. Besides, a photoluminescent measurement can be undertaken in various modalities such as lifetime and photoluminescent resonant-energy-transfer (PRET) measurement [72–74]. This versatile technique is adaptable to various situations.

Through the wide choices of photoluminescent probes and simultaneous optical excitation, photoluminescent detection is an ideal technique for multiplexed and high-throughput immunosensing, such as antibody microarrays. The photoluminescent intensity can be enhanced with metal nanostructures [39] and photonic crystals [40] fabricated on the sensor. Li *et al.* [41] reported a nanostructured aluminium-oxide (NAO)-based photoluminescence platform that enhanced a photoluminescent signal up to 100 fold and promoted the photoluminescent-based immunosensing with high throughput. Protein A and fluorophore-labeled immunoglobulin G (IgG) were used to demonstrate that the IgG of programmable concentrations was confirmed with the photoluminescent images on the microfluidic device. The sensitivity of the device as a concentration to detect IgG can be as small as 20 pg/mm^2 with a photoluminescent microscope. This photoluminescence-enhancing technology provides a powerful tool for photoluminescent-based immunosensing.

3. Label-Free Immunosensing

3.1. Change of Refractive Index

The optical detection methods mentioned above require either enzymatic or photoluminescent probes labeled on the detection antibody. Not only does the labeling impose additional cost and time, but also the structural modification might interfere with the binding of an antibody to an analyte. A label-free immunosensor is undoubtedly more economical and allows a detection in an unperturbed state. The label-free optical immunosensing typically exploits the local changes of refractive index (RI) induced with an antibody-antigen binding event [75,76]. The refractive indices of an aqueous solution and a protein are 1.33 and 1.5, respectively; the binding of an antibody on the sensor surface causes the local refractive index to shift from 1.33 to an effective refractive index in a range from 1.33 to 1.5 depending on the density of the antibodies bound on the surface. According to the Fresnel equation [77], the phase and amplitude of a reflected and transmitted beam at the interface depend on the refractive index and the thickness of each underlying layer. RI changes thus affect various optical phenomena such as reflection and interference at the interface, which can then serve as approaches to probe RI changes. Examples of these are ellipsometric [78] and interferometer-based immunosensors [79,80]. Another case is the existence of a particular resonant mode, which condition is sensitive to the interfacial RI changes. A surface-plasmon-resonance (SPR) immunosensor is a representative example and is examined in detail below. Other examples of resonance-based sensing include a ring resonator, resonant mirrors and metal-clad waveguide. Among all label-free immunosensors, an SPR-based biosensor has been commercialized and gained popularity in the biomedical research community. Although a label-free approach has labor-saving benefits, an adsorption of molecules other than the target antibodies cannot be discerned; the responses associated with non-binding events are the major sources of error for immunosensing of this type. For this reason, sample preprocessing and experimental controls are crucial steps in label-free immunosensing.

3.2. Surface-Plasmon-Resonance (SPR) Immunosensor

A surface plasmon (SP) is an electromagnetic resonant mode existing at a metal-dielectric interface [81–83]; this surface mode is closely related to the refractive indices of the two adjacent media. It carries a mathematic form similar to that of the evanescent wave; the generated electric field decays exponentially away from the interface (Figure 3). As a result, the sensing of a surface-plasmon resonance (SPR) is localized at the interface, excluding other reactions occurring in the bulk region. The concomitant evanescent wave with total reflection is typically employed to excite SP; this excitation requires the match of a wave vector (*k*) at a designated wavelength (λ). The wave vector of the evanescent wave can be modulated on adjusting the incident angle of the beam; when *k* matches that of the SP, a resonance occurs and light is absorbed and transformed to an SP. A standard SPR curve appears in Figure 3. The SPR sensor uses glass as a substrate and a thin layer of gold, typically 50 nm, is coated thereon. The SP lies at the gold-medium interface, and thus the sensing surface; upon the binding of antibodies, the waveform of SP is perturbed and the resonance curves then shift in response to the changes in SP. The reflectance change taken at a fixed angle can be an indicator of binding, proportional to the amount of analyte. Besides the planar SPR sensor, metallic nanoparticles exhibit SPR effects and can serve as immunosensors. The SP present at a metal-ambient interface absorbs light at a specific wavelength; the resonance frequency depends on the shape of the nanoparticle and the ambient refractive index [84]. The general design immobilizes an antibody on a gold nanoparticle; the binding of antibodies and an antigen initiates the aggregation of nanoparticles, which exhibit a surface-plasmon resonance at another frequency [85].

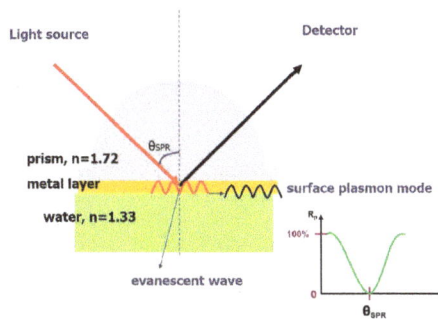

Figure 3. A prism-based SPR sensor with a gold film as sensing layer. The light is absorbed by surface plasmons at a particular resonant angle.

A planar SPR immunosensor can be expanded to an imaging mode (SPRI), which takes the reflection intensity as an indicator of binding and binding quantities. This simple setup has a capability of carrying highly sensitive immunosensing and high-throughput screening. It saves the labor of labeling an antibody with a photoluminescent probe; the antibodies can be regenerated for multiple use. A kinetic binding curve can be acquired in real time on an SPR immunosensor, which can be used to investigate the specific binding between an antibody and an antigen. Although SPRI is capable of incorporating multiple antibodies in parallel for detection, current commercial instruments are limited to a single analyte stream on a multiplexed sensor surface. Integration with microfluidic devices fully realizes the potential of SPRI in multiplexed and multi-analyte screening. Ouellet *et al.* [44] developed an integrated microfluidic array consisting of 264 addressable chambers with each isolated by microvalves and then connected by a serial network for simultaneous measurement of up to six analytes. Besides planar SPR sensors, colorimetric change from metallic nanoparticles is an exhibition of

SPR effects; this mechanism enables the design of a particle-based SPR sensor. Gold nanoparticles have long been used in a lateral flow device for rapid and disposable immunosensing [86]; the most popular example is a pregnancy test strip. Recent microfluidic designs further implement gold-nanoparticle sensing on newer materials such as cheap polyester cards [87] and chromatographic papers [88].

4. Microfluidics

Microfluidics offers an automatic way to dispense, to merge and to mix small volumes of samples for immunosensing [89,90]. The future development of immunosensing in a medical laboratory aims at rapid, sensitive and multiplex tests with a lancet blood sample. This condition requires sensitive optical detection with appropriate microfluidics integrated on the sensor. One clear goal of microfluidic immunosensing is to perform POC screening with the least support of other equipment. Power-free microfluidics is a promising tool for this purpose and has gained much attention from both academia and industry. Interesting designs of immunosensors using power-free and passive microfluidics, such as paper-based microfluidics and lateral flow devices, have been created, and are introduced in Section 4.1, following. For more accurate manipulation of a liquid sample, as required for a sensitive analysis in a medical laboratory, active and well controlled microfluidics are preferable. Popular designs in this category include Polydimethylsiloxane (PDMS) microchannels with micropneumatic or electrical pumping, compact disk driven by centrifugal force, and digital microfluidic techniques such as electrowetting-on-dielectric (EWOD). The EWOD technique especially performs well in multiple liquid processing, which includes dividing, separating, and transporting a tiny droplet for multiple detection with a simple setup, easily integrated with other optical components. Recent reports of the use of EWOD to manipulate an analyte in optical immunosensing are reviewed in the last Section 4.2.

4.1. Paper-Based Microfluidics (μPADs)

Paper, composed mainly of cellulose fiber, wicks an aqueous liquid and transports a liquid passively. It has long been used in chromatography to separate and to identify mixtures of small molecules, amino acids, proteins and antibodies [91]. Paper can be chemically modified with various functional groups that can covalently bind to proteins or DNA, or small molecules. Hydrophobic materials such as wax can be infused into sheets of paper as a barrier to create hydrophilic channels of desired patterns. The pattern and thickness define the dimension of the microfluidic channels on paper. The capillary property of cellulose fibers allows fluids to wick along the channels; there are some factors affecting the rate of wicking, which include the dimensions of the channel, the characteristics of the paper and the ambient temperature and humidity. Besides delivering flow through patterned microfluidic channels, the cellulose matrix of paper can be a solid support for liquid processing, such as filtering samples or performing chromatographic separations. Paper-based microchannels have several advantages over conventional PDMS microchannels as they are cheaper and easier to be fabricated. Paper-based devices are also biodegradable and can be mass-produced to meet the large-scale demand of POC screening.

Microfluidic paper-based analytical devices (μPADs) constitute POC diagnostic devices of a new class specifically designed for use in developing countries. A first paper-based ELISA was devised by Cheng *et al.* [92]; 96-well microzones, resembling a 96-well ELISA plate, were fabricated on paper using photolithographic processing; the hydrophobic photoresist layer confined the liquid reaction to the microzone. The main advantage of paper ELISA over a conventional plastic well is the saving of cost and time. In conventional methods, larger volumes of analytes and reagents (20–200 μL) are required; incubation and blocking duration for each step is long ($\geqslant 1$ h per step) because the analyte requires time to diffuse to the surface of the well. In comparison with the plastic well, paper offers a capillary support to deliver the analyte, thus decreasing the reaction duration. Each microzone requires only 3 μL reagent; the test is completed within an hour, as opposed to 2 h in conventional methods. The colorimetric result is measurable with a desktop scanner (cost $100), rather than an expensive microtitre plate reader. One disadvantage of paper-based ELISA is the decreased sensitivity, which is 54 fmol/zone,

one tenth of that obtained with an ELISA plate. This effect is possibly due to the decreased period of antibody-antigen incubation (lower loading) or non-specific interactions between antibodies and cellulose fibers (Figure 4).

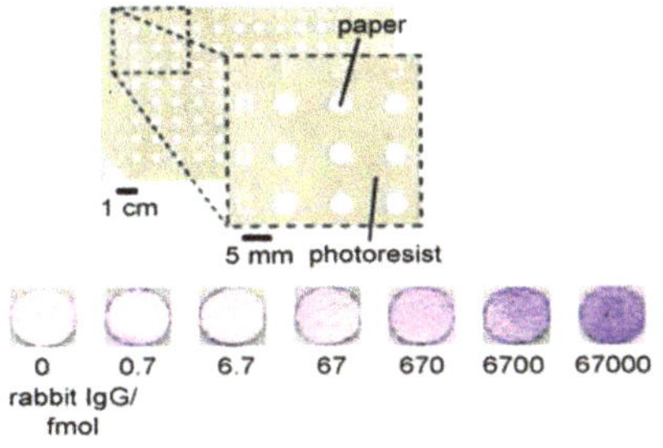

Figure 4. A 96-microzone paper-based ELISA [92]. Reproduction of the figures is made with permission of Wiley.

Paper-based ELISA expands the popular lateral-flow immunosensing to a two-dimensional multiplexed microfluidic immunosensing system. 3D paper-based devices have even more powerful capability, in which μL volumes of samples can be injected from a single inlet point into arrays of detection zones numbering up to thousands [93]. The fluids are distributed vertically and laterally; one stream travels across another without mixing. The 3D paper based microfluidic channels are created by stacking alternating layers of paper and water-impermeable double-sided adhesive tape. The stacking is patterned to allow the liquid flow within the layers of paper, but this design allows only colorimetric detection on the surface layers (top and bottom). Liu *et al.* [94] devised a 3D paper microfluidic device assembled with origami methods (see Figure 5). A piece of chromatography paper can be patterned with channels, reservoirs and frames in a single lithographic step; the paper is then folded along the frames in a sequence, which ensures that channels and reservoirs are aligned to build the 3D connection. Solutions are injected into four holes on the top. The paper is eventually unfolded so that all reservoirs are accessible for colorimetric analysis. Not only did Liu *et al.* propose a novel origami 3D design but also it is the first 3D paper-based immunosensor to adopt photoluminescent sensing, which detection typically offers greater sensitivity and smaller detection limits than simple colorimetric measurements. The photoluminescent detection is based on a dye, epicocconone, which exhibits enhanced photoluminescence in the presence of the analyte protein BSA (bovine serum albumin). Only 1.0 μL of a buffered epicocconone solution is required for the spotting on a detection reservoir; then the analyte BSA is injected from the inlet and reaches the reservoir to react. The layer is then scanned with a photoluminescent imager with resolution 100 μm within 1 min.

The primary detection methods for a paper-based immunosensor are still the qualitative colorimetric methods based on visual comparison with the naked eye or camera telephone. Quantitative analysis is important when the level of an analyte is clinically significant. Other methods such as chemiluminescence provide more precise and sensitive detection and concurrently retain simplicity, portability and modest cost of a paper device. Ge *et al.* [95] developed a paper-based ECL immunosensor on screen-printing carbon electrodes directly on paper; reservoir patterns were first created with wax printing. Carbon working electrodes were then screen-printed on a piece of square paper; another paper was printed with Ag/AgCl reference and carbon counter electrodes. The two papers were then aligned and stacked to form the reservoirs of the electrochemical cells, in total eight electrodes. A panel of biomarkers was screened with this design. With the aid of a chemiluminescent detector, eight working electrodes were sequentially triggered to produce ECL and detected. The linear dynamic range is 0.5–100 ng/mL. This device provides fresh opportunities for sensitive and precise paper-based immunosensors.

The possibility of label-free detection on a paper-based device can be envisaged from the work of Tseng *et al.* [96], involving a plasmonic sensor fabricated on paper using photothermal effects. Chemically synthesized gold nanoparticles have long been used as a sensing platform. This paper-based plasmonic gold-nanoparticle sensor was fabricated with an alternative method: a gold film was first deposited on a paper and laser-induced annealing was performed on the surface; the metal film absorbed the photon energy and converted it into thermal energy, which induced a local formation of nanoparticles. Using this nanostructure as a sensing surface with flow facilitated by the paper offers the possibility of a paper-based plasmonic immunosensor.

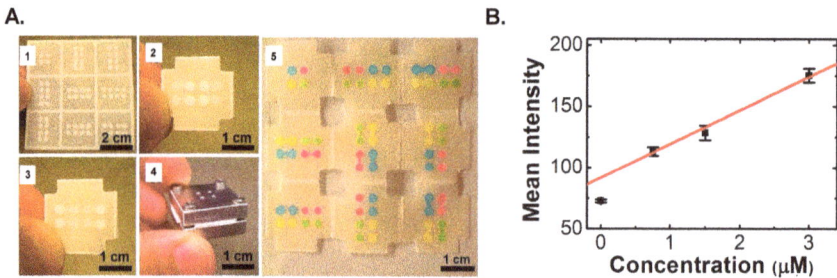

Figure 5. Paper-based microfluidics [94]. (**A**) A 3D paper-based immunosensor using an origami method: **1**. Photolithographically patterned channels on chromatography paper; **2**. Top layer of the device; **3**. Bottom layer of the device; **4**. Aluminium housing; **5**. Colored solutions were injected into designated channels. (**B**) Photoluminescent detection on the paper surface. Reproduction of the figures is made with permission of American Chemical Society.

4.2. Electrowetting-on-Dielectric (EWOD) Digital Microfluidic Devices

Digital microfluidics (DMF) is an emerging droplet-manipulation technique that depends on electric forces, such as electrowetting-on-dielectric (EWOD) or dielectrophoresis (DEP) [97,98]. These forces are caused by the potential difference between electrodes from an electrode array, which are coated with a dielectric and a hydrophobic insulator layer [99–102] on indium tin oxide (InSnO) glass or printed circuit boards (PCB). Electrowetting-on-dielectric (EWOD) digital microfluidic devices have been endorsed in many reports as a powerful platform for biological and biomedical research, including proteomic analysis [103–106], single-cell analysis [107–110], immunoassays [111], and clinical diagnostics [112].

On a DMF device, the small droplets can be individually operated to mix varied reagents, to incubate a reaction mixture, to split a droplet, to wash the modified surfaces or magnetic beads for immunosensing and finally to discard to a reservoir. Because each droplet is controlled with EWOD or DEP forces, complicated channels and external elements (such as pumps or valves or mechanical mixers) are unessential [113]. The comparative advantages of DMF over channel-based microfluidic devices include a simple device configuration, ease of modular interfaces for another integrating system, less requisite total sample volume for assay, less reagent consumption and a high potential for automation [109]. Much attention has thus been paid to the biological and biomedical application of DMF. Here we focus on immunosensors using DMF.

The immunosensing depends on a specific antibody-antigen interaction. In this process, multiple reagents with an antibody, a secondary antibody (conjugated with a tracer, such as HRP, photoluminescent dye or microparticles) and substrates (colorimetric or optical signal) are incubated, respectively; thorough washing with a washing buffer is essential. The DMF device has the advantages stated above to integrate the complicated processes on a device. The multiple droplets with varied reagents for immunosensing might be manipulated with EWOD or DEP forces through designed electrode arrays on the DMF device. The electrodes are coated with a hydrophobic insulator layer, such as Teflon-AF (amorphous fluoroplastics), for smooth operation of a droplet. Since direct immobilization of an antibody or antigen on a Teflon surface is difficult, immunosensing on a DMF device requires an additional solid-phase immobilization, such as magnetic beads [43,114] or surface modification to remove the surface hydrophobicity [115]. Sista *et al.* [43,114] used magnetic beads modified with antibodies to detect human insulin, interleukin (IL)-6 and troponin I from whole-blood samples on a DMF device. Vergauwe *et al.* [116] used magnetic beads to detect the presence of human IgE; the sensitivity of the device to detect IgE is at concentrations of about 150 nM. In another example, the device surface was directly modified; the antibodies were captured onto the hydrophobic surface of a DMF device. Miller *et al.* [115] introduced a surface-based immunoassay using DMF to detect the model analyte human IgG. Ng *et al.* [117] introduced novel magnetic particle-based immunosensing on a DMF device to detect thyroid stimulating hormone (TSH) and 17β-estradiol (E2). These powerful techniques have a great potential for the quantitative analysis of biomarks for diagnosis. The enzymatic chemiluminescent reaction was typically integrated on the DMF device [43,114,117] for optical detection. The photoluminescently labeled probes [115,116] and electrochemical detection [118] were also used. A label-free SPR technique was integrated on DMF devices [45,46]; Malic *et al.* revealed a dynamically configurable microarray SPR sensing on a DMF platform for real-time DNA hybridization. The application of SPR on a DMF platform might achieve a high-throughput screening process.

DMF has further integrated other techniques to perform immunoassays for sundry biomarkers. These DMF devices have great advantages to implement rapid and inexpensive instruments for diagnosis. Shamsi *et al.* [119] introduced a novel DMF device for magnetic microparticle-based immunosensing with a simple colorimetric detection to detect thyroid stimulating hormone (TSH); the sensitivity of this platform is 2.4 μIU/mL for the tested clinical applications. Rackus *et al.* [36] reported the first integration of nanostructured microelectrodes (NME) with a DMF platform to execute an electrochemical immunosensing for rubella virus (see Figure 6A); the sensitivity of the detection of this immunosensing is 0.07 IU/mL (5.7 μg/mL). Tsaloglou *et al.* [120] described a heterogeneous immunoassay with magnetic beads on a DMF platform to detect the cardiac marker Troponin I (cTnI); the sensitivity of the detection of this immunosensing is 2.0 ng/mL. Zeng *et al.* [62] reported the first experiment to integrate a portable prototype of a CL detector on a DMF platform (see Figure 6B); they manipulated the ball-like droplet to focus better the enzymatic chemiluminescence and thus enhanced the detection sensitivity of the optical signal. The detection limit was 0.01 mM H_2O_2. These integrated platforms shed light on highly sensitive CL detection for portable diagnostic devices [36,121,122].

Paper-based devices are mass-producible and disposable; they are suitable for rapid, single use and point-of-care screening. As opposed to paper-based microfluidic immunosensors, the EWOD-based device has more precise control of flow volume and can easily draw multiple droplets from the reservoir for multiplex sensing without elaborative reconstruction of sensor and system. DMF devices employing EWOD have more potential to be developed as desktop immunosensors with high sensitivity and very low sample volume. Compared to other microfluidic platforms, immunosensing using EWOD has not been fully explored yet and the currently reported LOD levels of EOWD-based immunosensors are not satisfactory (μg/mL level, Table 1). However, future development will incorporate other highly sensitive optical detection, such as ECL, in the EWOD-based system. As EWOD-based system has the advantages of simpler device configuration, easy integrating system and high potential for automation, it holds promise for future highly sensitive microfluidic immunosensors.

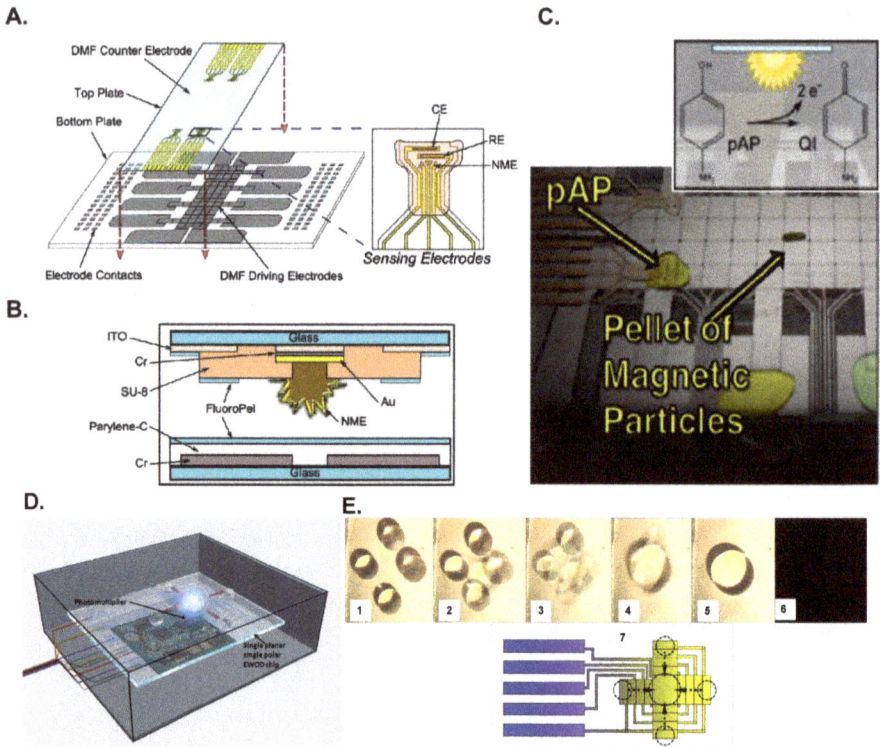

Figure 6. DMF further integrated another platform to perform immunoassays as described in: (**A–C**) Rackus *et al.* [36]; (**D,E**) Zeng *et al.* [62]. (**A**) DMF device with integrated nanostructured microelectrodes; (**B**) Schematic of a cross section of a DMF device; (**C**) Electrochemical measurements of the DMF device; (**D**) Schematic of the DMF chemiluminescent detector; (**E**) Mixing process on the DMF device (**1–5**); chemiluminescent photo (**6**) and schematic diagram (**7**). Reproduction of the figures is made with permission of Royal Society of Chemistry.

5. Conclusions

This review presents an overview of recent advances in opto-microfluidic immunosensing. Optical detection has long been the most popular detection technique for immunosensing. The advantages of the microfluidic immunosensor devices include decreased requirement of reagents and samples, small power consumption, small size, compact system, modest cost and a high potential for automation. The optical immunosensing process might be miniaturized and integrated with microfluidics to make compact lab-on-a-chip immunosensors. We believe that these integrated platforms of various opto-microfluidic immunosensor designs will fulfill future expectations for portable point-of-care immunosensing diagnostics.

Acknowledgments: Jie-Long He thanks Ministry of Science and Technology, Taiwan for providing a postdoctoral research scholarship (MOST 104-2218-E-002-025-MY3). Da-Shin Wang thanks Ministry of Science and Technology, Taiwan for providing a post-doctoral fellowship (MOST 104-2628-E-002-007-MY3).

Author Contributions: Shih-Kang Fan conceived and designed the structure of the review article; Jie-Long He and Da-Shin Wang collected the references and wrote the paper; they contributed equally to this work.

Conflicts of Interest: The specified sponsors had no role in the design of the study, in the collection or analyses or interpretation of data, in the writing of the manuscript, and in the decision to publish the results.

References

1. Rongen, H.A.; Hoetelmans, R.M.; Bult, A.; van Bennekom, W.P. Chemiluminescence and immunoassays. *J. Pharm. Biomed. Anal.* **1994**, *12*, 433–462. [CrossRef]
2. Singh, P. Dendrimers and their applications in immunoassays and clinical diagnostics. *Biotechnol. Appl. Biochem.* **2007**, *48*, 1–9. [CrossRef] [PubMed]
3. Dhawan, S. Signal amplification systems in immunoassays: Implications for clinical diagnostics. *Expert Rev. Mol. Diagn.* **2006**, *6*, 749–760. [CrossRef] [PubMed]
4. Wheeler, M.J. Immunoassay techniques. *Methods Mol. Biol.* **2006**, *324*, 1–23. [PubMed]
5. Afeyan, N.B.; Gordon, N.F.; Regnier, F.E. Automated real-time immunoassay of biomolecules. *Nature* **1992**, *358*, 603–604. [CrossRef] [PubMed]
6. Chen, Y.T.; Juang, R.H.; He, J.L.; Chu, W.Y.; Wang, C.H. Detection of H6 influenza antibody by blocking enzyme-linked immunosorbent assay. *Vet. Microbiol.* **2010**, *142*, 205–210. [CrossRef] [PubMed]
7. He, J.L.; Hsieh, M.S.; Chiu, Y.C.; Juang, R.H.; Wang, C.H. Preparation of monoclonal antibodies against poor immunogenic avian influenza virus proteins. *J. Immunol. Methods* **2013**, *387*, 43–50. [CrossRef] [PubMed]
8. He, J.L.; Hsieh, M.S.; Juang, R.H.; Wang, C.H. A monoclonal antibody recognizes a highly conserved neutralizing epitope on hemagglutinin of H6N1 avian influenza virus. *Vet. Microbiol.* **2014**, *174*, 333–341. [CrossRef] [PubMed]
9. Nagatani, N.; Yamanaka, K.; Ushijima, H.; Koketsu, R.; Sasaki, T.; Ikuta, K.; Saito, M.; Miyahara, T.; Tamiya, E. Detection of influenza virus using a lateral flow immunoassay for amplified DNA by a microfluidic RT-PCR chip. *Analyst* **2012**, *137*, 3422–3426. [CrossRef] [PubMed]
10. MacCallum, R.M.; Martin, A.C.; Thornton, J.M. Antibody-antigen interactions: Contact analysis and binding site topography. *J. Mol. Biol.* **1996**, *262*, 732–745. [CrossRef] [PubMed]
11. Porstmann, T.; Kiessig, S.T. Enzyme immunoassay techniques an overview. *J. Immunol. Methods* **1992**, *150*, 5–21. [CrossRef]
12. Bange, A.; Halsall, H.B.; Heineman, W.R. Microfluidic immunosensor systems. *Biosens. Bioelectron.* **2005**, *20*, 2488–2503. [CrossRef] [PubMed]
13. Choi, S.; Goryll, M.; Sin, L.Y.M.; Wong, P.K.; Chae, J. Microfluidic-based biosensors toward point-of-care detection of nucleic acids and proteins. *Microfluid. Nanofluid.* **2011**, *10*, 231–247. [CrossRef]
14. Henares, T.G.; Mizutani, F.; Hisamoto, H. Current development in microfluidic immunosensing chip. *Anal. Chim. Acta* **2008**, *611*, 17–30. [CrossRef] [PubMed]
15. Lin, C.-C.; Wang, J.-H.; Wu, H.-W.; Lee, G.-B. Microfluidic immunoassays. *J. Lab. Autom.* **2010**, *15*, 253–274. [CrossRef]
16. Hey, T.; Fiedler, E.; Rudolph, R.; Fiedler, M. Artificial, non-antibody binding proteins for pharmaceutical and industrial applications. *Trends Biotechnol.* **2005**, *23*, 514–522. [CrossRef] [PubMed]
17. Jayasena, S.D. Aptamers: An emerging class of molecules that rival antibodies in diagnostics. *Clin. Chem.* **1999**, *45*, 1628–1650. [PubMed]
18. Makaraviciute, A.; Ramanaviciene, A. Site-directed antibody immobilization techniques for immunosensors. *Biosens. Bioelectron.* **2013**, *50*, 460–471. [CrossRef] [PubMed]
19. Peluso, P.; Wilson, D.S.; Do, D.; Tran, H.; Venkatasubbaiah, M.; Quincy, D.; Heidecker, B.; Poindexter, K.; Tolani, N.; Phelan, M.; *et al.* Optimizing antibody immobilization strategies for the construction of protein microarrays. *Anal. Biochem.* **2003**, *312*, 113–124. [CrossRef]
20. Vashist, S.K.; Dixit, C.K.; MacCraith, B.D.; O'Kennedy, R. Effect of antibody immobilization strategies on the analytical performance of a surface plasmon resonance-based immunoassay. *Analyst* **2011**, *136*, 4431–4436. [CrossRef] [PubMed]
21. Corry, B.; Uilk, J.; Crawley, C. Probing direct binding affinity in electrochemical antibody-based sensors. *Anal. Chim. Acta* **2003**, *496*, 103–116. [CrossRef]
22. Wang, J. Electrochemical biosensors: Towards point-of-care cancer diagnostics. *Biosens. Bioelectron.* **2006**, *21*, 1887–1892. [CrossRef] [PubMed]
23. Zang, D.; Ge, L.; Yan, M.; Song, X.; Yu, J. Electrochemical immunoassay on a 3D microfluidic paper-based device. *Chem. Commun.* **2012**, *48*, 4683–4685. [CrossRef] [PubMed]

24. Fodey, T.L.; Thompson, C.S.; Traynor, I.M.; Haughey, S.A.; Kennedy, D.G.; Crooks, S.R. Development of an optical biosensor based immunoassay to screen infant formula milk samples for adulteration with melamine. *Anal. Chem.* **2011**, *83*, 5012–5016. [CrossRef] [PubMed]

25. Mauriz, E.; Calle, A.; Montoya, A.; Lechuga, L.M. Determination of environmental organic pollutants with a portable optical immunosensor. *Talanta* **2006**, *69*, 359–364. [CrossRef] [PubMed]

26. Xu, S. Electromechanical biosensors for pathogen detection. *Microchim. Acta* **2012**, *178*, 245–260. [CrossRef]

27. Mecklenburg, M.; Lindbladh, C.; Li, H.; Mosbach, K.; Danielsson, B. Enzymatic amplification of a flow-injected thermometric enzyme-linked immunoassay for human insulin. *Anal. Biochem.* **1993**, *212*, 388–393. [CrossRef] [PubMed]

28. Novo, P.; Prazeres, D.M.; Chu, V.; Conde, J.P. Microspot-based ELISA in microfluidics: Chemiluminescence and colorimetry detection using integrated thin-film hydrogenated amorphous silicon photodiodes. *Lab Chip* **2011**, *11*, 4063–4071. [CrossRef] [PubMed]

29. Yu, L.; Li, C.M.; Liu, Y.; Gao, J.; Wang, W.; Gan, Y. Flow-through functionalized PDMS microfluidic channels with dextran derivative for ELISAs. *Lab Chip* **2009**, *9*, 1243–1247. [CrossRef] [PubMed]

30. Xu, H.; Mao, X.; Zeng, Q.; Wang, S.; Kawde, A.N.; Liu, G. Aptamer-functionalized gold nanoparticles as probes in a dry-reagent strip biosensor for protein analysis. *Anal. Chem.* **2009**, *81*, 669–675. [CrossRef] [PubMed]

31. Lee, B.S.; Lee, J.N.; Park, J.M.; Lee, J.G.; Kim, S.; Cho, Y.K.; Ko, C. A fully automated immunoassay from whole blood on a disc. *Lab Chip* **2009**, *9*, 1548–1555. [CrossRef] [PubMed]

32. Pires, N.M.M.; Dong, T. Ultrasensitive opto-microfluidic immunosensor integrating gold nanoparticle-enhanced chemiluminescence and highly stable organic photodetector. *J. Biomed. Opt.* **2014**, *19*, 030504. [CrossRef] [PubMed]

33. Yao, P.; Liu, Z.; Tung, S.; Dong, Z.; Liu, L. Fully automated quantification of insulin concentration using a microfluidic-based chemiluminescence immunoassay. *J. Lab. Autom.* **2015**. [CrossRef] [PubMed]

34. Ge, L.; Wang, S.; Song, X.; Ge, S.; Yu, J. 3D origami-based multifunction-integrated immunodevice: Low-cost and multiplexed sandwich chemiluminescence immunoassay on microfluidic paper-based analytical device. *Lab Chip* **2012**, *12*, 3150–3158. [CrossRef] [PubMed]

35. Guo, S.; Ishimatsu, R.; Nakano, K.; Imato, T. Automated chemiluminescence immunoassay for a nonionic surfactant using a recycled spinning-pausing controlled washing procedure on a compact disc-type microfluidic platform. *Talanta* **2015**, *133*, 100–106. [CrossRef] [PubMed]

36. Rackus, D.G.; Dryden, M.D.; Lamanna, J.; Zaragoza, A.; Lam, B.; Kelley, S.O.; Wheeler, A.R. A digital microfluidic device with integrated nanostructured microelectrodes for electrochemical immunoassays. *Lab Chip* **2015**, *15*, 3776–3784. [CrossRef] [PubMed]

37. Wang, S.; Ge, L.; Zhang, Y.; Song, X.; Li, N.; Ge, S.; Yu, J. Battery-triggered microfluidic paper-based multiplex electrochemiluminescence immunodevice based on potential-resolution strategy. *Lab Chip* **2012**, *12*, 4489–4498. [CrossRef] [PubMed]

38. Wu, M.S.; Shi, H.W.; He, L.J.; Xu, J.J.; Chen, H.Y. Microchip device with 64-site electrode array for multiplexed immunoassay of cell surface antigens based on electrochemiluminescence resonance energy transfer. *Anal. Chem.* **2012**, *84*, 4207–4213. [CrossRef] [PubMed]

39. Sang, C.H.; Chou, S.J.; Pan, F.M.; Sheu, J.T. Fluorescence enhancement and multiple protein detection in ZnO nanostructure microfluidic devices. *Biosens. Bioelectron.* **2016**, *75*, 285–292. [CrossRef] [PubMed]

40. Tan, Y.F.; Tang, T.T.; Xu, H.S.; Zhu, C.Q.; Cunningham, B.T. High sensitivity automated multiplexed immunoassays using photonic crystal enhanced fluorescence microfluidic system. *Biosens. Bioelectron.* **2015**, *73*, 32–40. [CrossRef] [PubMed]

41. Li, X.; Yin, H.C.; Que, L. A nanostructured aluminum oxide-based microfluidic device for enhancing immunoassay's fluorescence and detection sensitivity. *Biomed. Microdevices* **2014**, *16*, 771–777. [CrossRef] [PubMed]

42. Rowe, C.A.; Scruggs, S.B.; Feldstein, M.J.; Golden, J.P.; Ligler, F.S. An array immunosensor for simultaneous detection of clinical analytes. *Anal. Chem.* **1999**, *71*, 433–439. [CrossRef] [PubMed]

43. Sista, R.S.; Hua, Z.; Thwar, P.; Sudarsan, A.; Srinivasan, V.; Eckhardt, A.E.; Pollack, M.G.; Pamula, V.K. Development of a digital microfluidic platform for point of care testing. *Lab Chip* **2008**, *8*, 2091–2104. [CrossRef] [PubMed]

44. Ouellet, E.; Lausted, C.; Lin, T.; Yang, C.W.T.; Hood, L.; Lagally, E.T. Parallel microfluidic surface plasmon resonance imaging arrays. *Lab Chip* **2010**, *10*, 581–588. [CrossRef] [PubMed]

45. Malic, L.; Veres, T.; Tabrizian, M. Biochip functionalization using electrowetting-on-dielectric digital microfluidics for surface plasmon resonance imaging detection of DNA hybridization. *Biosens. Bioelectron.* **2009**, *24*, 2218–2224. [CrossRef] [PubMed]

46. Malic, L.; Veres, T.; Tabrizian, M. Two-dimensional droplet-based surface plasmon resonance imaging using electrowetting-on-dielectric microfluidics. *Lab Chip* **2009**, *9*, 473–475. [CrossRef] [PubMed]

47. Kuswandi, B.; Nuriman; Huskens, J.; Verboom, W. Optical sensing systems for microfluidic devices: A review. *Anal. Chim. Acta* **2007**, *601*, 141–155. [CrossRef] [PubMed]

48. Engvall, E.; Jonsson, K.; Perlmann, P. Enzyme-linked immunosorbent assay. II. Quantitative assay of protein antigen, immunoglobulin G, by means of enzyme-labelled antigen and antibody-coated tubes. *Biochim. Biophys. Acta Protein Struct.* **1971**, *251*, 427–434. [CrossRef]

49. Lequin, R.M. Enzyme immunoassay (EIA)/enzyme-linked immunosorbent assay (ELISA). *Clin. Chem.* **2005**, *51*, 2415–2418. [CrossRef] [PubMed]

50. Schroeder, H.; Adler, M.; Gerigk, K.; Muller-Chorus, B.; Gotz, F.; Niemeyer, C.M. User configurable microfluidic device for multiplexed immunoassays based on DNA-directed assembly. *Anal. Chem.* **2009**, *81*, 1275–1279. [CrossRef] [PubMed]

51. Madou, M.; Zoval, J.; Jia, G.; Kido, H.; Kim, J.; Kim, N. Lab on a CD. *Annu. Rev. Biomed. Eng.* **2006**, *8*, 601–628. [CrossRef] [PubMed]

52. Dodeigne, C.; Thunus, L.; Lejeune, R. Chemiluminescence as diagnostic tool. A review. *Talanta* **2000**, *51*, 415–439. [CrossRef]

53. Zhao, L.X.; Sun, L.; Chu, X.G. Chemiluminescence immunoassay. *Trends Anal. Chem.* **2009**, *28*, 404–415. [CrossRef]

54. Fan, A.; Cao, Z.; Li, H.; Kai, M.; Lu, J. Chemiluminescence platforms in immunoassay and DNA analyses. *Anal. Sci.* **2009**, *25*, 587–597. [CrossRef] [PubMed]

55. Yakovleva, J.; Davidsson, R.; Lobanova, A.; Bengtsson, M.; Eremin, S.; Laurell, T.; Emneus, J. Microfluidic enzyme immunoassay using silicon microchip with immobilized antibodies and chemiluminescence detection. *Anal. Chem.* **2002**, *74*, 2994–3004. [CrossRef] [PubMed]

56. Baeyens, W.R.; Schulman, S.G.; Calokerinos, A.C.; Zhao, Y.; Garcia Campana, A.M.; Nakashima, K.; de Keukeleire, D. Chemiluminescence-based detection: Principles and analytical applications in flowing streams and in immunoassays. *J. Pharm. Biomed. Anal.* **1998**, *17*, 941–953. [CrossRef]

57. Radi, R.; Cosgrove, T.P.; Beckman, J.S.; Freeman, B.A. Peroxynitrite-induced luminol chemiluminescence. *Biochem. J.* **1993**, *290*, 51–57. [CrossRef] [PubMed]

58. Yang, Z.; Liu, H.; Zong, C.; Yan, F.; Ju, H. Automated support-resolution strategy for a one-way chemiluminescent multiplex immunoassay. *Anal. Chem.* **2009**, *81*, 5484–5489. [CrossRef] [PubMed]

59. Yacoub-George, E.; Hell, W.; Meixner, L.; Wenninger, F.; Bock, K.; Lindner, P.; Wolf, H.; Kloth, T.; Feller, K.A. Automated 10-channel capillary chip immunodetector for biological agents detection. *Biosens. Bioelectron.* **2007**, *22*, 1368–1375. [CrossRef] [PubMed]

60. Wang, P.P.; Ge, L.; Yan, M.; Song, X.R.; Ge, S.G.; Yu, J.H. Paper-based three-dimensional electrochemical immunodevice based on multi-walled carbon nanotubes functionalized paper for sensitive point-of-care testing. *Biosens. Bioelectron.* **2012**, *32*, 238–243. [CrossRef] [PubMed]

61. Weeks, I.; Sturgess, M.L.; Woodhead, J.S. Chemiluminescence immunoassay: An overview. *Clin. Sci.* **1986**, *70*, 403–408. [CrossRef] [PubMed]

62. Zeng, X.; Zhang, K.; Pan, J.; Chen, G.; Liu, A.Q.; Fan, S.K.; Zhou, J. Chemiluminescence detector based on a single planar transparent digital microfluidic device. *Lab Chip* **2013**, *13*, 2714–2720. [CrossRef] [PubMed]

63. Chemical Connection. Available online: http://www.chemicalconnection.org.uk/chemistry/topics/images/ee7b.jpg (accessed on 5 Febuary 2016).

64. Hu, L.; Xu, G. Applications and trends in electrochemiluminescence. *Chem. Soc. Rev.* **2010**, *39*, 3275–3304. [CrossRef] [PubMed]

65. Richter, M.M. Electrochemiluminescence (ECL). *Chem. Rev.* **2004**, *104*, 3003–3036. [CrossRef] [PubMed]

66. Forster, R.J.; Bertoncello, P.; Keyes, T.E. Electrogenerated chemiluminescence. *Annu. Rev. Anal. Chem.* **2009**, *2*, 359–385. [CrossRef] [PubMed]

67. De Lorimier, R.M.; Smith, J.J.; Dwyer, M.A.; Looger, L.L.; Sali, K.M.; Paavola, C.D.; Rizk, S.S.; Sadigov, S.; Conrad, D.W.; Loew, L.; *et al.* Construction of a fluorescent biosensor family. *Protein Sci.* **2002**, *11*, 2655–2675. [CrossRef] [PubMed]
68. Melnyk, O.; Duburcq, X.; Olivier, C.; Urbès, F.; Auriault, C.; Gras-Masse, H. Peptide arrays for highly sensitive and specific antibody-binding fluorescence assays. *Bioconjug. Chem.* **2002**, *13*, 713–720. [CrossRef] [PubMed]
69. Cho, I.H.; Mauer, L.; Irudayaraj, J. *In-situ* fluorescent immunomagnetic multiplex detection of foodborne pathogens in very low numbers. *Biosens. Bioelectron.* **2014**, *57*, 143–148. [CrossRef] [PubMed]
70. De Beéck, K.O.; Vermeersch, P.; Verschueren, P.; Westhovens, R.; Mariën, G.; Blockmans, D.; Bossuyt, X. Antinuclear antibody detection by automated multiplex immunoassay in untreated patients at the time of diagnosis. *Autoimmun. Rev.* **2012**, *12*, 137–143. [CrossRef] [PubMed]
71. Juncker, D.; Bergeron, S.; Laforte, V.; Li, H. Cross-reactivity in antibody microarrays and multiplexed sandwich assays: Shedding light on the dark side of multiplexing. *Curr. Opin. Chem. Biol.* **2014**, *18*, 29–37. [CrossRef] [PubMed]
72. Hou, J.Y.; Liu, T.C.; Lin, G.F.; Li, Z.X.; Zou, L.P.; Li, M.; Wu, Y.S. Development of an immunomagnetic bead-based time-resolved fluorescence immunoassay for rapid determination of levels of carcinoembryonic antigen in human serum. *Anal. Chim. Acta* **2012**, *734*, 93–98. [CrossRef] [PubMed]
73. Liu, Y.; Zhou, S.; Tu, D.; Chen, Z.; Huang, M.; Zhu, H.; Ma, E.; Chen, X. Amine-functionalized lanthanide-doped zirconia nanoparticles: Optical spectroscopy, time-resolved fluorescence resonance energy transfer biodetection, and targeted imaging. *J. Am. Chem. Soc.* **2012**, *134*, 15083–15090. [CrossRef] [PubMed]
74. Wang, Q.; Nchimi Nono, K.; Syrjänpää, M.; Charbonnière, L.J.; Hovinen, J.; Härmä, H. Stable and highly fluorescent europium (III) chelates for time-resolved immunoassays. *Inorg. Chem.* **2013**, *52*, 8461–8466. [CrossRef] [PubMed]
75. Lin, B.; Qiu, J.; Gerstenmeier, J.; Li, P.; Pien, H.; Pepper, J.; Cunningham, B. A label-free optical technique for detecting small molecule interactions. *Biosens. Bioelectron.* **2002**, *17*, 827–834. [CrossRef]
76. Fan, X.; White, I.M.; Shopova, S.I.; Zhu, H.; Suter, J.D.; Sun, Y. Sensitive optical biosensors for unlabeled targets: A review. *Anal. Chim. Acta* **2008**, *620*, 8–26. [CrossRef] [PubMed]
77. Bennett, C.A. *Principles of Physical Optics*; Wiley: Hoboken, NJ, USA, 2008.
78. Jin, G.; Meng, Y.H.; Liu, L.; Niu, Y.; Chen, S.; Cai, Q.; Jiang, T.J. Development of biosensor based on imaging ellipsometry and biomedical applications. *Thin Solid Films* **2011**, *519*, 2750–2757. [CrossRef]
79. Alvarez, S.D.; Li, C.P.; Chiang, C.E.; Schuller, I.K.; Sailor, M.J. A label-free porous alumina interferometric immunosensor. *ACS Nano* **2009**, *3*, 3301–3307. [CrossRef] [PubMed]
80. Mun, K.S.; Alvarez, S.D.; Choi, W.Y.; Sailor, M.J. A stable, label-free optical interferometric biosensor based on TiO$_2$ nanotube arrays. *ACS Nano* **2010**, *4*, 2070–2076. [CrossRef] [PubMed]
81. Homola, J.; Yee, S.S.; Gauglitz, G. Surface plasmon resonance sensors: Review. *Sens. Actuators B Chem.* **1999**, *54*, 3–15. [CrossRef]
82. Pattnaik, P. Surface plasmon resonance: Applications in understanding receptor-ligand interaction. *Appl. Biochem. Biotechnol.* **2005**, *126*, 79–92. [CrossRef]
83. Kanda, V.; Kariuki, J.K.; Harrison, D.J.; McDermott, M.T. Label-free reading of microarray-based immunoassays with surface plasmon resonance imaging. *Anal. Chem.* **2004**, *76*, 7257–7262. [CrossRef] [PubMed]
84. Willets, K.A.; van Duyne, R.P. Localized surface plasmon resonance spectroscopy and sensing. *Annu. Rev. Phys. Chem.* **2007**, *58*, 267–297. [CrossRef] [PubMed]
85. Thanh, N.T.K.; Rosenzweig, Z. Development of an aggregation-based immunoassay for anti-protein a using gold nanoparticles. *Anal. Chem.* **2002**, *74*, 1624–1628. [CrossRef] [PubMed]
86. Posthuma-Trumpie, G.A.; Korf, J.; van Amerongen, A. Lateral flow (immuno) assay: Its strengths, weaknesses, opportunities and threats. A literature survey. *Anal. Bioanal. Chem.* **2009**, *393*, 569–582. [CrossRef] [PubMed]
87. Lafleur, L.; Stevens, D.; McKenzie, K.; Ramachandran, S.; Spicar-Mihalic, P.; Singhal, M.; Arjyal, A.; Osborn, J.; Kauffman, P.; Yager, P. Progress toward multiplexed sample-to-result detection in low resource settings using microfluidic immunoassay cards. *Lab Chip* **2012**, *12*, 1119–1127. [CrossRef] [PubMed]

88. Fu, E.; Liang, T.; Spicar-Mihalic, P.; Houghtaling, J.; Ramachandran, S.; Yager, P. Two-dimensional paper network format that enables simple multistep assays for use in low-resource settings in the context of malaria antigen detection. *Anal. Chem.* **2012**, *84*, 4574–4579. [CrossRef] [PubMed]

89. Ng, A.H.; Uddayasankar, U.; Wheeler, A.R. Immunoassays in microfluidic systems. *Anal. Bioanal. Chem.* **2010**, *397*, 991–1007. [CrossRef] [PubMed]

90. Foudeh, A.M.; Fatanat Didar, T.; Veres, T.; Tabrizian, M. Microfluidic designs and techniques using lab-on-a-chip devices for pathogen detection for point-of-care diagnostics. *Lab Chip* **2012**, *12*, 3249–3266. [CrossRef] [PubMed]

91. Zhao, W.; van der Berg, A. Lab on paper. *Lab Chip* **2008**, *8*, 1988–1991. [PubMed]

92. Cheng, C.M.; Martinez, A.W.; Gong, J.; Mace, C.R.; Phillips, S.T.; Carrilho, E.; Mirica, K.A.; Whitesides, G.M. Paper-based ELISA. *Angew. Chem. Int. Ed. Engl.* **2010**, *49*, 4771–4774. [CrossRef] [PubMed]

93. Martinez, A.W.; Phillips, S.T.; Whitesides, G.M. Three-dimensional microfluidic devices fabricated in layered paper and tape. *Proc. Natl. Acad. Sci. USA* **2008**, *105*, 19606–19611. [CrossRef] [PubMed]

94. Liu, H.; Crooks, R.M. Three-dimensional paper microfluidic devices assembled using the principles of origami. *J. Am. Chem. Soc.* **2011**, *133*, 17564–17566. [CrossRef] [PubMed]

95. Ge, L.; Yan, J.X.; Song, X.R.; Yan, M.; Ge, S.G.; Yu, J.H. Three-dimensional paper-based electrochemiluminescence immunodevice for multiplexed measurement of biomarkers and point-of-care testing. *Biomaterials* **2012**, *33*, 1024–1031. [CrossRef] [PubMed]

96. Tseng, S.C.; Yu, C.C.; Wan, D.; Chen, H.L.; Wang, L.A.; Wu, M.C.; Su, W.F.; Han, H.C.; Chen, L.C. Eco-friendly plasmonic sensors: Using the photothermal effect to prepare metal nanoparticle-containing test papers for highly sensitive colorimetric detection. *Anal. Chem.* **2012**, *84*, 5140–5145. [CrossRef] [PubMed]

97. Fan, S.K.; Hsieh, T.H.; Lin, D.Y. General digital microfluidic platform manipulating dielectric and conductive droplets by dielectrophoresis and electrowetting. *Lab Chip* **2009**, *9*, 1236–1242. [CrossRef] [PubMed]

98. Fan, S.K.; Hsu, Y.W.; Chen, C.H. Encapsulated droplets with metered and removable oil shells by electrowetting and dielectrophoresis. *Lab Chip* **2011**, *11*, 2500–2508. [CrossRef] [PubMed]

99. Shih, S.C.; Gach, P.C.; Sustarich, J.; Simmons, B.A.; Adams, P.D.; Singh, S.; Singh, A.K. A droplet-to-digital (D2D) microfluidic device for single cell assays. *Lab Chip* **2015**, *15*, 225–236. [CrossRef] [PubMed]

100. Rival, A.; Jary, D.; Delattre, C.; Fouillet, Y.; Castellan, G.; Bellemin-Comte, A.; Gidrol, X. An EWOD-based microfluidic chip for single-cell isolation, mRNA purification and subsequent multiplex qPCR. *Lab Chip* **2014**, *14*, 3739–3749. [CrossRef] [PubMed]

101. Fan, S.K.; Huang, P.W.; Wang, T.T.; Peng, Y.H. Cross-scale electric manipulations of cells and droplets by frequency-modulated dielectrophoresis and electrowetting. *Lab Chip* **2008**, *8*, 1325–1331. [CrossRef] [PubMed]

102. Fan, S.K.; Yang, H.; Wang, T.T.; Hsu, W. Asymmetric electrowetting—Moving droplets by a square wave. *Lab Chip* **2007**, *7*, 1330–1335. [CrossRef] [PubMed]

103. Wheeler, A.R.; Moon, H.; Kim, C.J.; Loo, J.A.; Garrell, R.L. Electrowetting-based microfluidics for analysis of peptides and proteins by matrix-assisted laser desorption/ionization mass spectrometry. *Anal. Chem.* **2004**, *76*, 4833–4838. [CrossRef] [PubMed]

104. Kirby, A.E.; Lafreniere, N.M.; Seale, B.; Hendricks, P.I.; Cooks, R.G.; Wheeler, A.R. Analysis on the go: Quantitation of drugs of abuse in dried urine with digital microfluidics and miniature mass spectrometry. *Anal. Chem.* **2014**, *86*, 6121–6129. [CrossRef] [PubMed]

105. Wheeler, A.R.; Moon, H.; Bird, C.A.; Loo, R.R.; Kim, C.J.; Loo, J.A.; Garrell, R.L. Digital microfluidics with in-line sample purification for proteomics analyses with MALDI-MS. *Anal. Chem.* **2005**, *77*, 534–540. [CrossRef] [PubMed]

106. Mei, N.; Seale, B.; Ng, A.H.; Wheeler, A.R.; Oleschuk, R. Digital microfluidic platform for human plasma protein depletion. *Anal. Chem.* **2014**, *86*, 8466–8472. [CrossRef] [PubMed]

107. Sanders, R.; Huggett, J.F.; Bushell, C.A.; Cowen, S.; Scott, D.J.; Foy, C.A. Evaluation of digital PCR for absolute DNA quantification. *Anal Chem* **2011**, *83*, 6474–6484. [CrossRef] [PubMed]

108. White, R.A., III; Quake, S.R.; Curr, K. Digital PCR provides absolute quantitation of viral load for an occult RNA virus. *J. Virol. Methods* **2012**, *179*, 45–50. [CrossRef] [PubMed]

109. He, J.L.; Chen, A.T.; Lee, J.H.; Fan, S.K. Digital microfluidics for manipulation and analysis of a single cell. *Int. J. Mol. Sci.* **2015**, *16*, 22319–22332. [CrossRef] [PubMed]

110. Barbulovic-Nad, I.; Yang, H.; Park, P.S.; Wheeler, A.R. Digital microfluidics for cell-based assays. *Lab Chip* **2008**, *8*, 519–526. [CrossRef] [PubMed]

111. Yetisen, A.K.; Akram, M.S.; Lowe, C.R. Paper-based microfluidic point-of-care diagnostic devices. *Lab Chip* **2013**, *13*, 2210–2251. [CrossRef] [PubMed]

112. Pollack, M.G.; Pamula, V.K.; Srinivasan, V.; Eckhardt, A.E. Applications of electrowetting-based digital microfluidics in clinical diagnostics. *Expert Rev. Mol. Diagn.* **2011**, *11*, 393–407. [CrossRef] [PubMed]

113. Choi, K.; Ng, A.H.; Fobel, R.; Wheeler, A.R. Digital microfluidics. *Annu. Rev. Anal. Chem.* **2012**, *5*, 413–440. [CrossRef] [PubMed]

114. Sista, R.S.; Eckhardt, A.E.; Srinivasan, V.; Pollack, M.G.; Palanki, S.; Pamula, V.K. Heterogeneous immunoassays using magnetic beads on a digital microfluidic platform. *Lab Chip* **2008**, *8*, 2188–2196. [CrossRef] [PubMed]

115. Miller, E.M.; Ng, A.H.; Uddayasankar, U.; Wheeler, A.R. A digital microfluidic approach to heterogeneous immunoassays. *Anal. Bioanal. Chem.* **2011**, *399*, 337–345. [CrossRef] [PubMed]

116. Vergauwe, N.; Witters, D.; Ceyssens, F.; Vermeir, S.; Verbruggen, B.; Puers, R.; Lammertyn, J. A versatile electrowetting-based digital microfluidic platform for quantitative homogeneous and heterogeneous bio-assays. *J. Micromech. Microeng.* **2011**, *21*, 054026. [CrossRef]

117. Ng, A.H.; Choi, K.; Luoma, R.P.; Robinson, J.M.; Wheeler, A.R. Digital microfluidic magnetic separation for particle-based immunoassays. *Anal. Chem.* **2012**, *84*, 8805–8812. [CrossRef] [PubMed]

118. Karuwan, C.; Sukthang, K.; Wisitsoraat, A.; Phokharatkul, D.; Patthanasettakul, V.; Wechsatol, W.; Tuantranont, A. Electrochemical detection on electrowetting-on-dielectric digital microfluidic chip. *Talanta* **2011**, *84*, 1384–1389. [CrossRef] [PubMed]

119. Shamsi, M.H.; Choi, K.; Ng, A.H.; Wheeler, A.R. A digital microfluidic electrochemical immunoassay. *Lab Chip* **2014**, *14*, 547–554. [CrossRef] [PubMed]

120. Tsaloglou, M.N.; Jacobs, A.; Morgan, H. A fluorogenic heterogeneous immunoassay for cardiac muscle troponin cTnI on a digital microfluidic device. *Anal. Bioanal. Chem.* **2014**, *406*, 5967–5976. [CrossRef] [PubMed]

121. Ng, A.H.; Lee, M.; Choi, K.; Fischer, A.T.; Robinson, J.M.; Wheeler, A.R. Digital microfluidic platform for the detection of rubella infection and immunity: A proof of concept. *Clin. Chem.* **2015**, *61*, 420–429. [CrossRef] [PubMed]

122. Choi, K.; Ng, A.H.; Fobel, R.; Chang-Yen, D.A.; Yarnell, L.E.; Pearson, E.L.; Oleksak, C.M.; Fischer, A.T.; Luoma, R.P.; Robinson, J.M.; *et al.* Automated digital microfluidic platform for magnetic-particle-based immunoassays with optimization by design of experiments. *Anal. Chem.* **2013**, *85*, 9638–9646. [CrossRef] [PubMed]

micromachines

MDPI

Review

Constriction Channel Based Single-Cell Mechanical Property Characterization [†]

Chengcheng Xue [1], Junbo Wang [1,*], Yang Zhao [1], Deyong Chen [1], Wentao Yue [2,*] and Jian Chen [1,*]

[1] State Key Laboratory of Transducer Technology, Institute of Electronics, Chinese Academy of Sciences, Beijing 100190, China; xuechengcheng13@mails.ucas.ac.cn (C.X.); zhaoyang110@mails.ucas.ac.cn (Y.Z.); dychen@mail.ie.ac.cn (D.C.)

[2] Department of Cellular and Molecular Biology, Beijing Chest Hospital, Capital Medical University, Beijing101149, China

* Correspondence: jbwang@mail.ie.ac.cn (J.W.); yuewt@ccmu.edu.cn (W.Y.); chenjian@mail.ie.ac.cn (J.C.); Tel./Fax: +86-10-5888-7191 (J.W.); +86-10-8950-9373 (W.Y.); +86-10-5888-7531 (ext. 816) (J.C.)

[†] This paper is an extended version of our paper published in the 5th International Conference on Optofluidics 2015, Taipei, Taiwan, 26–29 July 2015.

Academic Editors: Shih-Kang Fan, Da-Jeng Yao and Yi-Chung Tung
Received: 2 September 2015 ; Accepted: 10 November 2015 ; Published: 16 November 2015

Abstract: This mini-review presents recent progresses in the development of microfluidic constriction channels enabling high-throughput mechanical property characterization of single cells. We first summarized the applications of the constriction channel design in quantifying mechanical properties of various types of cells including red blood cells, white blood cells, and tumor cells. Then we highlighted the efforts in modeling the cellular entry process into the constriction channel, enabling the translation of raw mechanical data (e.g., cellular entry time into the constriction channel) into intrinsic cellular mechanical properties such as cortical tension or Young's modulus. In the end, current limitations and future research opportunities of the microfluidic constriction channels were discussed.

Keywords: single-cell analysis; microfluidics; constriction channel; mechanical property characterization; high throughput

1. Introduction

The mechanical properties of a biological cell are largely determined by the characteristics of its cytoskeleton, an elaborate network of fibrous proteins [1,2]. Various diseases and changes in cell states are reported to lead to variations in cellular mechanical properties, which include (1) changes in the stiffness of blood cells (e.g., variations of red blood cells (RBCs) in malaria or sickle cell anemia and white blood cells (WBCs) in sepsis, trauma, and acute respiratory distress syndrome); (2) increased cell deformability of invasive cancer cells; and (3) decreased deformability during the stem cell differentiation process [3,4].

Conventionally, cellular mechanical properties are quantified based on well-established techniques (e.g., atomic force microscopy, micropipette aspiration and optical tweezers) [3]. In atomic force microscopy, a probe tip attached to a flexible cantilever is pressed into the cell surface for a set distance with cantilever deflections measured and translated to cellular mechanical properties [5,6]. Although powerful, this approach is capable of quantifying cellular mechanical properties, which are dependent on experimental conditions (e.g., poking depth, rate, and position) [7]. To deal with this issue, multiple scans on the same cell are requested to collect trustworthy data, leading to a compromised measurement speed with limited throughput (e.g., less than 10 cells per sample from patient pleural fluids based on atomic force microscopy [8]).

For micropipette aspiration, the surface of a cell is aspirated into a small glass tube with the leading edge of its surface tracked, which is further translated to cellular mechanical properties [9]. Compared with atomic force microscopy, micropipette aspiration deforms a cell patch in a more global manner, leading to more accurate characterization of cellular mechanical properties. However, this technique requires skilled operation on the glass pipette and thus proceeds with limited throughput (~10 cells per cell type from patient voided urine [10,11]).

Microfluidics is the manipulation and processing of fluidics on micrometer scales ranging from one to hundreds of μms [12,13]. Due to its dimensional match with biological cells, it has been used for single-cell analysis [14,15]. Currently, several microfluidic devices have been proposed to quantify the mechanical properties of single cells based on various mechanisms such as optical stretching, fluid stretching and constriction channels [16,17].

In a microfluidic optical stretcher, a two-beam laser is used to serially deform single suspended cells flowing within microfluidic channels for cellular mechanical property characterization [18,19]. In a microfluidic hydrodynamic stretcher, single cells are delivered to a micro channel with geometry variations, producing extensional fluid flow to cause cell deformation [20–22]. Although these two approaches are featured with significantly higher throughputs than conventional approaches, they can only collect cellular mechanical parameters (e.g., deformation ratio and elongation index), which remain dependent on cell sizes and experimental conditions (e.g., pressure drop, channel geometry) [17].

Table 1. Key developments in the field of microfluidic constriction channels enabling high-throughput cellular mechanical property characterization.

Cell Types	Quantified Parameters	Key Observations	References
Plasmodium falciparum infected RBCs	Channel blockage	Maturation of Plasmodium falciparum decreased the deformability of infected RBCs.	[23]
Plasmodium vivax infected RBCs	Channel blockage	No significant decrease in deformability was observed during the maturation of Plasmodium vivax infected RBCs.	[27]
Plasmodium falciparum infected RBCs	Transit velocity	The parasite protein Pf155 decreased the transit velocity of ring-stage infected RBCs.	[29]
Plasmodium falciparum infected RBCs	Cortical tension	Cortical tensions were quantified as 3.22 ± 0.64 pN/μm (uninfected RBCs), 4.66 ± 1.15 pN/μm (ring-stage infected RBCs), 8.26 ± 2.84 pN/μm (early trophozoite infected RBCs), and 21.38 ± 5.81 pN/μm (late trophozoite infected RBCs).	[30]
Plasmodium falciparum infected RBCs	Transit velocity	Artesunate (a drug in malaria) decreased the deformability of infected RBCs while it had no effect on normal RBCs.	[31]
Normal and oxidized RBCs	Cortical tension	Cortical tensions were quantified as 20.13 ± 1.47 pN/μm (normal RBCs) and 27.51 ± 3.64 pN/μm (oxidized RBCs).	[32]
White blood cells	Transit time	In diseases of sepsis and leukostasis, decreases in the deformability of WBCs were found.	[24]
Breast tumor cells	Entry time and transit velocity	Benign breast epithelial cells of MCF-10A had longer entry times than tumor breast cells of MCF-7 with similar sizes.	[25]
Lung tumor cells	Instantaneous Young's modulus	Instantaneous Young's moduli were quantified as 3.48 ± 0.86 kPa for A549 cells and 2.99 ± 0.38 kPa for 95C cells	[47]

Meanwhile, the microfluidic constriction channel is used to quantify the cellular entry and transition process through a micro channel with a cross-sectional area smaller than the dimensions of a single cell, enabling high-throughput single-cell mechanical property characterization [23–26] (see Table 1). This technique was first used to evaluate the mechanical properties of RBCs [23,27–40], which was then expanded to study the deformability of WBCs [24,41,42] and tumor cells [25,43–45]. Leveraging mechanical modeling of the cellular entry process into the constriction channel, the microfluidic constriction channel design can collect size-independent intrinsic biomechanical markers such as cortical tension or Young's modulus [35,46–48].

2. Constriction Channel Based Mechanical Property Characterization of Red Blood Cells

Initially, the constriction channel design was used to quantify the mechanical properties of RBCs [23,27–40]. In 2003, Chiu *et al.*, characterized complex behaviors of *Plasmodium falciparum* infected RBCs using the constriction channels with sizes at 8, 6, 4, and 2 μm in width [23] (see Figure 1a). Ring-stage infected RBCs resembled normal RBCs in morphology and were capable of passing through all constricted channels. Early and late trophozoite infected RBCs were noticed squeezing through the larger 8- and 6-μm channels but would block the smaller 4- and 2-μm channels. Schizont stage infected RBCs blocked all but the 8-μm channels (see Figure 1b). In addition, the same constriction channel design with 2-μm channels was used to probe the mechanical properties of *Plasmodium vivax* infected RBCs, revealing that in contrast to *Plasmodium falciparum*, *Plasmodium vivax* infected RBCs of all developmental stages were observed to transverse the 2-μm constriction readily [27].

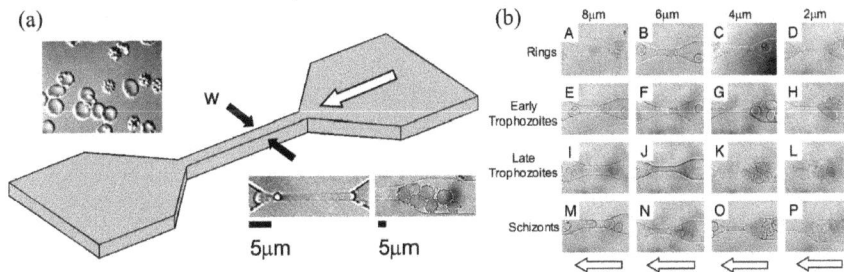

Figure 1. (**a**) Schematic illustration of the geometry of the constriction channel design which was sized at 8, 6, 4, and 2 μm in width. (Upper Inset) An image of normal (smooth) and infected RBCs. (Lower Inset) A normal RBC was passing through a 2-μm constriction and infected RBCs blocked a 6-μm constriction channel. (**b**) Four stages of malaria-infected RBCs through the constriction channels. Ring-stage infected erythrocytes were able to pass through all constriction channels. Early trophozoite and late trophozoite infected cells passed through the larger 8- and 6-μm channels but eventually blocked the smaller 4- and 2-μm channels. Schizont stage infected erythrocytes blocked all but the 8-μm channels. The arrows indicate direction of flow. Reproduced with permission from [23].

Using microfluidic constriction channels in parallel, the effect of Artesunate (a drug widely used for the treatment of malaria) on the dynamic deformability of RBCs infected with ring-stage *Plasmodium falciparum* malaria was also evaluated [31]. As shown in Figure 2a, a microfluidic device with triangular pillar arrays as repeated constrictions (inter-pillar gap sizes of 3 or 4 μm) was used and RBCs were deformed to pass through constriction channels in series (see Figure 2b). After the treatment of Artesunate, a 50% decrease in the transit velocity of *Plasmodium falciparum* infected RBCs was reported whereas only small (~10%) velocity reduction was observed among uninfected RBCs. These results demonstrated that ART alters the deformability of *Plasmodium falciparum* infected RBCs, which may influence blood circulation through micro vasculatures (see Figure 2c).

Furthermore, Ma *et al.*, accurately controlled the pressure applied across the constriction channels to obtain the critical pressure threshold requested to push each RBCs through the constriction positions. Based on an equivalent mechanical model, the threshold value was translated to an intrinsic mechanical parameter, cortical tension (see Figure 3a) [30]. This approach was used to study the mechanical properties of *Plasmodium falciparum* infected RBCs, producing cortical tensions of 3.22 ± 0.64 pN/μm for uninfected RBCs, 4.66 ± 1.15 pN/μm for ring-stage infected RBCs, 8.26 ± 2.84 pN/μm for early trophozoite infected RBCs, and 21.38 ± 5.81 pN/μm for late trophozoite infected RBCs. In addition, the measured cortical tensions of schizont stage infected RBCs ranged from 85 to 1300 pN/μm, with an average value of 606 pN/μm (see Figure 3b).

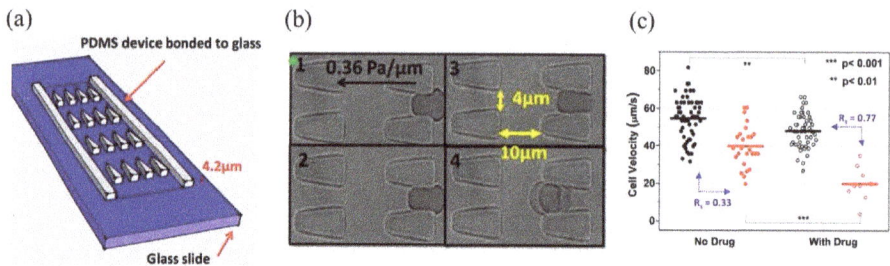

Figure 2. (**a**) Schematic of the geometry of the parallel constriction channel design; and (**b**) experimental images of RBCs passing through constriction positions; (**c**) The treatment of Artesunate significantly decreases the cell velocity of *Plasmodium falciparum* infected RBCs while it has no significant effect on uninfected RBCs. Reproduced with permission from [31].

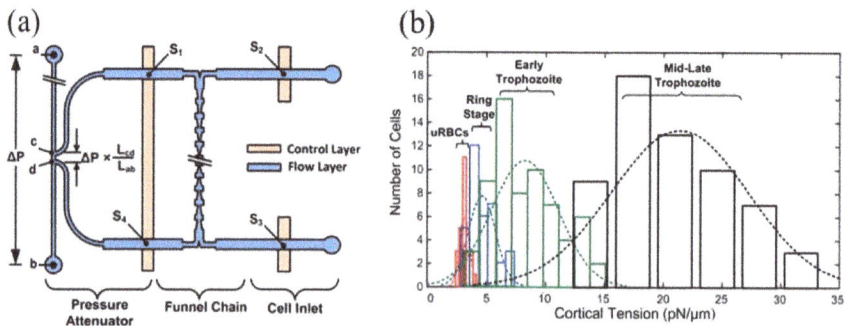

Figure 3. (**a**) Design of the flow and control layers of the microfluidic constriction channel capable of generating precisely controlled pressure to quantify the critical threshold requested to push each RBC through the constriction channel. In this study, the pressure regulator divides an externally applied pressure by a factor of 100, which is further applied across the funnel chain; (**b**) Histogram of the quantified cortical tensions of RBCs in various stages. Reproduced with permission from [30].

3. Constriction Channel Based Mechanical Property Characterization of White Blood Cells

The constriction channel design was also used to characterize the mechanical properties of WBCs [24,41,42,49]. Flectcher *et al.*, were pioneers in this field and proposed a network of constriction channels (~6 μm wide) to characterize transit time of WBCs in diseases of sepsis and leukostasis (see Figure 4a) [24]. Experimental results show that (1) inflammatory mediators involved in sepsis significantly increased the transit time of WBCs (see Figure 4b); (2) altered mechanical properties of WBCs were found to correlate with symptoms of leukostasis in patients.

Furthermore, Theodoly *et al.*, explored the functions of actin organization and myosin II leveraging the constriction channel design, revealing that (1) cell stiffness depends strongly on the organization of F-actin rather than myosin II; (2) the actin network is not completely destroyed after a forced travelling through the constriction channel; (3) myosin II plays a major role in maintaining cellular shapes [41].

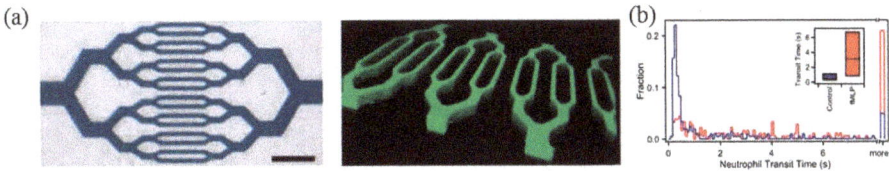

Figure 4. (**a**) Schematic of the constriction channel network for the mechanical property characterization of WBCs where the device trifurcates into a network of bifurcating channels including 64 parallel constriction channels; (**b**) The transit time of the neutrophil populations with or without fMLP exposure. Note that fMLP is an inflammatory mediator involved in sepsis. Reproduced with permission from [24].

4. Constriction Channel Based Mechanical Property Characterization of Tumor Cells

The constriction channel design was further used to characterize and classify tumor cells based on their mechanical properties [25,43–45,50]. In 2008, Lim *et al.*, classified benign breast epithelial cells (MCF-10A) and non-metastatic tumor breast cells (MCF-7) using the constriction channel design, finding that MCF-10A had longer entry time than MCF-7 with similar sizes [25]. In addition, Vanapalli *et al.*, classified benign and cancerous brain cells using the constriction channels, revealing that compared to the cellular transit velocity within the constriction channels, cellular entry time into the constriction channel can provide more insights in differentiating these cell types [43]. Furthermore, Erickson *et al.*, proposed a microfluidic device with serial constriction channels to study the repeated deformability of tumor cells with and without the treatment of taxol. Experimental results show that (1) cells treated with taxol required longer transit times when travelling through the first constriction than untreated cells; (2) the initial transit required the longest time and subsequent transits were faster and the difference between the two cell groups was reduced (see Figure 5) [44].

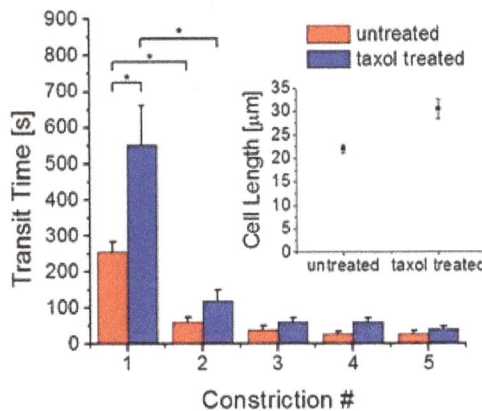

Figure 5. Cell permeation across sequential microfluidic constriction channels with the effects of taxol investigated. Cells treated with taxol were larger (inset) and required a longer transit time to cross the first constriction than untreated cells. For both cell groups, the initial transit required the longest time while subsequent transits were faster where the duration difference between the two cell groups was reduced. Reproduced with permission from [44].

5. Mechanical Modeling of the Constriction Channel Design

In the majority of the aforementioned studies, raw mechanical parameters including entry time, transit velocity and elongation index were commonly derived from the constriction channel design for cell type classification. However, these parameters are strongly dependent on cell sizes and experimental conditions (e.g., dimensions of the constriction channels and pressure applied to push cells through the constriction channels). Thus, they cannot reflect intrinsic cellular mechanical properties of single cells.

To address this issue, a few studies have been conducted to model the cellular entry process into the constriction channel which can translate these raw parameters into intrinsic cellular mechanical parameters such as Young's Modulus and Cortical Tension [35,36,45–47,51,52]. Ma *et al.*, are pioneers in this field and they used a Newtonian liquid drop to model a RBC where the cell deformability is indicated by a persistent cortical tension of the cell membrane (see Figure 6) [35,46]. When each cell is constrained in a constriction channel, the cellular deformation can be divided into three sections: a leading portion, an internal section contacting the constriction, and a trailing portion (see Figure 6a). Based on this equivalent mechanical model, the quantified critical threshold pressure which can push an RBC through the constriction channel was transferred to cortical tensions. Based on this approach, the cortical tensions of neutrophils, lymphocytes, RT4 bladder cancer cells, and lymphoma cells were quantified as 37.0 ± 4.8, 74.7 ± 9.8, 185.4 ± 25.3, and 235.4 ± 31.0 pN/µm, respectively (see Figure 6b).

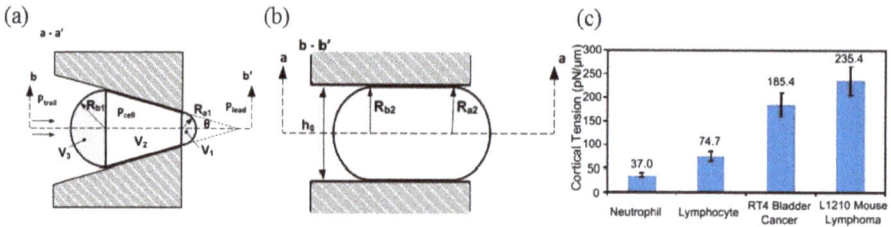

Figure 6. (**a**) Top view; and (**b**) side view illustration of a single RBC at the critical point of the constriction constriction; (**c**) Comparison of the quantified cortical tensions from four different cell types. Reproduced with permission from [46].

In addition, Theodoly *et al.*, modelled the cellular entry process into the constriction channel with the effects of friction and leakage taken into consideration [51]. As a follow-up study, Chen *et al.*, used a visco-hyperelastic solid to model a tumor cell using ABAQUS based numerical simulations where cell-channel wall frictions were modeled using a simple Coulomb law (see Figure 7) [47]. Both experiments and simulations confirmed the two-stage cellular entry process into the constriction channel: an instantaneous jump into the channel indicated by instantaneous aspiration length followed by a creeping increase in aspiration length which is terminated by transitional aspiration length (see Figure 7a). Numerical simulations reveal that instantaneous aspiration length and transitional aspiration length are reversely proportional to Young's instantaneous modulus ($E_{\text{instantaneous}}$), which are affected by friction coefficient regardless of other cellular viscoelastic parameters. By combining measured instantaneous aspiration length and transitional aspiration length with these obtained from numerical simulations, $E_{\text{instantaneous}}$ of these two cells types were quantified as 3.48 ± 0.86 kPa (A549 cells, $n = 199$) and 2.99 ± 0.38 kPa (95C cells, $n = 164$) (see Figure 7b).

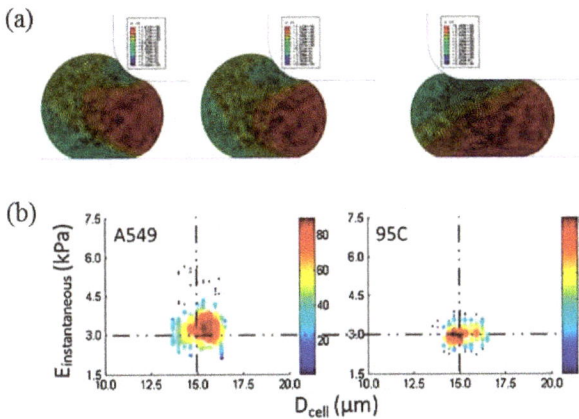

Figure 7. (a) Numerical simulations of the cellular entry process including initial jump into the channel (left), cellular creep response (middle) and creep response termination (right) at the transitional position. (b) Scatter plots of $E_{instantaneous}$ (Instantaneous Young's Modulus) *vs.* D_{cell} for A549 cells and 95C cells (two lung cancer cell lines), indicating noticeable differences in $E_{instantaneous}$ between these two cell types. Reproduced with permission from [47].

6. Discussions

Due to dimensional comparison, the microfluidic constriction channel is a powerful tool in the field of single-cell analysis, which was used for quantifying cellular mechanical properties where single cells were flushed into the constriction channel with their entry times and transition velocities adopted as biophysical markers [23–25]. Furthermore, this tool was used for quantifying cellular electrical properties since the deformed cells effectively seal constriction channel walls, and block electric lines, enabling single-cell electrical property characterization [53–56]. Meanwhile, the constriction channel design was also used as a tool to study chemical synthesis of red blood cells [57] or deliver vector-free gene vectors [42,58].

In the field of constriction channel based single-cell mechanical property characterization, after roughly 10 years of intensive studies, two challenges have been carefully addressed. The first key progress is the capability of translating raw parameters such as entry time into intrinsic mechanical markers such as instantaneous Young's modulus or cortical tension [35,46,47]. These raw mechanical parameters depend on cell sizes and experimental conditions. Thus, data reported by different groups cannot be effectively compared. Meanwhile, intrinsic mechanical parameters such as Young's modulus and cortical tension are not affected by these environmental changes, and thus data collected from multiple centers can be collected and compared, which may help form detection thresholds in classifying corresponding diseases. Note that the up-to-date understanding of cells' mechanical properties considers cells as a viscoelastic material and a comprehensive determination of a cell's mechanical properties requires additional data such as relaxation time besides Young's modulus or cortical tensions.

The second key progress is the proposal of constriction channels with adjustable cross-section areas to deal with the issue of channel blockage [59,60]. Since the dimensions of the constriction channel are smaller than single cells, there is a persistent concern of channel clogging due to unwanted microparticles. In recent studies, one fixed wall of the constriction channel was replaced with a thin membrane and its deflection was regulated by external pressure sources. If the constriction channel was blocked by solid particles or cell clusters, the application of pressure causes the deflection of the constriction channel walls, enlarging the cross-sectional area to remove the blocked particles. This improvement in the constriction channel design can significantly improve the robustness of the device in the process of characterizing cellular mechanical properties [59,60].

Currently, the major limitation of the constriction channel based single-cell mechanical property characterization is the throughput, which is roughly one cell per second. Although this value is much higher than conventional techniques such as atomic force microscopy and micropipette aspiration (one cell per min), it is significantly lower than the throughput of flow cytometry at roughly 1000 cells per second and thus cannot collect data from thousands of single cells.

The second concern is whether mechanical markers alone are enough to classify cell types. For diseases with blood cells or tumors, there are heterogeneous cell types and it is highly doubtful that biomechanical markers alone can provide a sound evaluation. Currently, the constriction channel design was integrated with impedance spectroscopy to enable both electrical and mechanical property characterization of single cells [61–63]. Further studies may consider integrating the constriction channel design with other functional components to further enable the measurement of multiple parameters of single cells, providing a comprehensive evaluation of biological cells.

Acknowledgments: We thank the National Basic Research Program of China (973 Program, Grant No. 2014CB744600), the National Natural Science Foundation of China (Grant No. 61201077, 61431019 and 81261120561), and the Beijing NOVA Program of Science and Technology for financial support.

Author Contributions: C.X. reviewed the section of mechanical property characterization of red blood cells. J.W. reviewed the section of mechanical property characterization of white blood cells. Y.Z. reviewed the section of mechanical property characterization of tumor cells. D.C. reviewed the section of mechanical modeling of the constriction channel. W.Y. contributed to the section of discussion. C.X. and J.C. drafted the manuscript.

References

1. Ethier, C.R.; Simmons, C.A. *Introductory Biomechanics: From Cells to Organisms*; Cambridge University Press: Cambridge, UK, 2007.
2. Fletcher, D.A.; Mullins, R.D. Cell mechanics and the cytoskeleton. *Nature* **2010**, *463*, 485–492. [CrossRef] [PubMed]
3. Di Carlo, D. A mechanical biomarker of cell state in medicine. *J. Lab. Autom.* **2012**, *17*, 32–42. [CrossRef] [PubMed]
4. Lee, G.Y.H.; Lim, C.T. Biomechanics approaches to studying human diseases. *Trends Biotechnol.* **2007**, *25*, 111–118. [CrossRef] [PubMed]
5. Alonso, J.L.; Goldmann, W.H. Feeling the forces: Atomic force microscopy in cell biology. *Life Sci.* **2003**, *72*, 2553–2560. [CrossRef]
6. Kirmizis, D.; Logothetidis, S. Atomic force microscopy probing in the measurement of cell mechanics. *Int. J. Nanomed.* **2010**, *5*, 137–145. [CrossRef]
7. Lekka, M.; Pogoda, K.; Gostek, J.; Klymenko, O.; Prauzner-Bechcicki, S.; Wiltowska-Zuber, J.; Jaczewska, J.; Lekki, J.; Stachura, Z. Cancer cell recognition–mechanical phenotype. *Micron* **2012**, *43*, 1259–1266. [CrossRef] [PubMed]
8. Cross, S.E.; Jin, Y.S.; Rao, J.; Gimzewski, J.K. Nanomechanical analysis of cells from cancer patients. *Nat. Nanotechnol.* **2007**, *2*, 780–783. [CrossRef] [PubMed]
9. Hochmuth, R.M. Micropipette aspiration of living cells. *J. Biomech.* **2000**, *33*, 15–22. [CrossRef]
10. Shojaei-Baghini, E.; Zheng, Y.; Jewett, M.A.S.; Geddie, W.B.; Sun, Y. Mechanical characterization of benign and malignant urothelial cells from voided urine. *Appl. Phys. Lett.* **2013**, *102*, 123704. [CrossRef]
11. Shojaei-Baghini, E.; Zheng, Y.; Sun, Y. Automated micropipette aspiration of single cells. *Ann. Biomed. Eng.* **2013**, *41*, 1208–1216. [CrossRef] [PubMed]
12. Whitesides, G.M. The origins and the future of microfluidics. *Nature* **2006**, *442*, 368–373. [CrossRef] [PubMed]
13. Wootton, R.C.; Demello, A.J. Microfluidics: Exploiting elephants in the room. *Nature* **2010**, *464*, 839–840. [CrossRef] [PubMed]

14. Weaver, W.M.; Tseng, P.; Kunze, A.; Masaeli, M.; Chung, A.J.; Dudani, J.S.; Kittur, H.; Kulkarni, R.P.; Di Carlo, D. Advances in high-throughput single-cell microtechnologies. *Curr. Opin. Biotechnol.* **2014**, *25*, 114–123. [CrossRef] [PubMed]

15. Sims, C.E.; Allbritton, N.L. Analysis of single mammalian cells on-chip. *Lab Chip* **2007**, *7*, 423–440. [CrossRef] [PubMed]

16. Kim, D.H.; Wong, P.K.; Park, J.; Levchenko, A.; Sun, Y. Microengineered platforms for cell mechanobiology. *Annu. Rev. Biomed. Eng.* **2009**, *11*, 203–233. [CrossRef] [PubMed]

17. Zheng, Y.; Nguyen, J.; Wei, Y.; Sun, Y. Recent advances in microfluidic techniques for single-cell biophysical characterization. *Lab Chip* **2013**, *13*, 2464–2483. [CrossRef] [PubMed]

18. Lincoln, B.; Erickson, H.M.; Schinkinger, S.; Wottawah, F.; Mitchell, D.; Ulvick, S.; Bilby, C.; Guck, J. Deformability-based flow cytometry. *Cytom. Part A* **2004**, *59*, 203–209. [CrossRef] [PubMed]

19. Guck, J.; Schinkinger, S.; Lincoln, B.; Wottawah, F.; Ebert, S.; Romeyke, M.; Lenz, D.; Erickson, H.M.; Ananthakrishnan, R.; Mitchell, D.; *et al.* Optical deformability as an inherent cell marker for testing malignant transformation and metastatic competence. *Biophys. J.* **2005**, *88*, 3689–3698. [CrossRef] [PubMed]

20. Gossett, D.R.; Tse, H.T.; Lee, S.A.; Ying, Y.; Lindgren, A.G.; Yang, O.O.; Rao, J.; Clark, A.T.; Di Carlo, D. Hydrodynamic stretching of single cells for large population mechanical phenotyping. *Proc. Natl. Acad. Sci. USA* **2012**, *109*, 7630–7635. [CrossRef] [PubMed]

21. Yaginuma, T.; Oliveira, M.S.; Lima, R.; Ishikawa, T.; Yamaguchi, T. Human red blood cell behavior under homogeneous extensional flow in a hyperbolic-shaped microchannel. *Biomicrofluidics* **2013**, *7*, 054110. [CrossRef] [PubMed]

22. Otto, O.; Rosendahl, P.; Mietke, A.; Golfier, S.; Herold, C.; Klaue, D.; Girardo, S.; Pagliara, S.; Ekpenyong, A.; Jacobi, A.; *et al.* Real-time deformability cytometry: On-the-fly cell mechanical phenotyping. *Nat. Meth.* **2015**, *12*, 199–202. [CrossRef] [PubMed]

23. Shelby, J.P.; White, J.; Ganesan, K.; Rathod, P.K.; Chiu, D.T. A microfluidic model for single-cell capillary obstruction by Plasmodium falciparum infected erythrocytes. *Proc. Natl. Acad. Sci. USA* **2003**, *100*, 14618–14622. [CrossRef] [PubMed]

24. Rosenbluth, M.J.; Lam, W.A.; Fletcher, D.A. Analyzing cell mechanics in hematologic diseases with microfluidic biophysical flow cytometry. *Lab Chip* **2008**, *8*, 1062–1070. [CrossRef] [PubMed]

25. Hou, H.W.; Li, Q.S.; Lee, G.Y.H.; Kumar, A.P.; Ong, C.N.; Lim, C.T. Deformability study of breast cancer cells using microfluidics. *Biomed. Microdevices* **2009**, *11*, 557–564. [CrossRef] [PubMed]

26. Abkarian, M.; Faivre, M.; Stone, H.A. High-speed microfluidic differential manometer for cellular-scale hydrodynamics. *Proc. Natl. Acad. Sci. USA* **2006**, *103*, 538–542. [CrossRef] [PubMed]

27. Handayani, S.; Chiu, D.T.; Tjitra, E.; Kuo, J.S.; Lampah, D.; Kenangalem, E.; Renia, L.; Snounou, G.; Price, R.N.; Anstey, N.M.; *et al.* High deformability of Plasmodium vivax-infected red blood cells under microfluidic conditions. *J. Infect. Dis.* **2009**, *199*, 445–450. [CrossRef] [PubMed]

28. Quinn, D.J.; Pivkin, I.; Wong, S.Y.; Chiam, K.H.; Dao, M.; Karniadakis, G.E.; Suresh, S. Combined simulation and experimental study of large deformation of red blood cells in microfluidic systems. *Ann. Biomed. Eng.* **2011**, *39*, 1041–1050. [CrossRef] [PubMed]

29. Diez-Silva, M.; Park, Y.; Huang, S.; Bow, H.; Mercereau-Puijalon, O.; Deplaine, G.; Lavazec, C.; Perrot, S.; Bonnefoy, S.; Feld, M.S.; *et al.* Pf155/RESA protein influences the dynamic microcirculatory behavior of ring-stage Plasmodium falciparum infected red blood cells. *Sci. Rep.* **2012**, *2*, 614. [CrossRef] [PubMed]

30. Guo, Q.; Reiling, S.J.; Rohrbach, P.; Ma, H. Microfluidic biomechanical assay for red blood cells parasitized by Plasmodium falciparum. *Lab Chip* **2012**, *12*, 1143–1150. [CrossRef] [PubMed]

31. Huang, S.; Undisz, A.; Diez-Silva, M.; Bow, H.; Dao, M.; Han, J. Dynamic deformability of Plasmodium falciparum-infected erythrocytes exposed to artesunate *in vitro*. *Integr. Biol.* **2013**, *5*, 414–422. [CrossRef] [PubMed]

32. Kwan, J.M.; Guo, Q.; Kyluik-Price, D.L.; Ma, H.; Scott, M.D. Microfluidic analysis of cellular deformability of normal and oxidatively damaged red blood cells. *Am. J. Hematol.* **2013**, *88*, 682–689. [CrossRef] [PubMed]

33. Wu, T.; Feng, J.J. Simulation of malaria-infected red blood cells in microfluidic channels: Passage and blockage. *Biomicrofluidics* **2013**, *7*, 44115. [CrossRef] [PubMed]

34. Sakuma, S.; Kuroda, K.; Tsai, C.D.; Fukui, W.; Arai, F.; Kaneko, M. Red blood cell fatigue evaluation based on the close-encountering point between extensibility and recoverability. *Lab Chip* **2014**, *14*, 1135–1141. [CrossRef] [PubMed]

35. Guo, Q.; Duffy, S.P.; Matthews, K.; Santoso, A.T.; Scott, M.D.; Ma, H. Microfluidic analysis of red blood cell deformability. *J. Biomech.* **2014**, *47*, 1767–1776. [CrossRef] [PubMed]
36. Myrand-Lapierre, M.-E.; Deng, X.; Ang, R.R.; Matthews, K.; Santoso, A.T.; Ma, H. Multiplexed fluidic plunger mechanism for the measurement of red blood cell deformability. *Lab Chip* **2015**, *15*, 159–167. [CrossRef] [PubMed]
37. Herricks, T.; Antia, M.; Rathod, P.K. Deformability limits of Plasmodium falciparum-infected red blood cells. *Cell. Microbiol.* **2009**, *11*, 1340–1353. [CrossRef] [PubMed]
38. Bow, H.; Pivkin, I.V.; Diez-Silva, M.; Goldfless, S.J.; Dao, M.; Niles, J.C.; Suresh, S.; Han, J. A microfabricated deformability-based flow cytometer with application to malaria. *Lab Chip* **2011**, *11*, 1065–1073. [CrossRef] [PubMed]
39. Wood, D.K.; Soriano, A.; Mahadevan, L.; Higgins, J.M.; Bhatia, S.N. A biophysical indicator of vaso-occlusive risk in sickle cell disease. *Sci. Transl. Med.* **2012**, *4*, 123ra26. [CrossRef] [PubMed]
40. Zeng, N.F.; Ristenpart, W.D. Mechanical response of red blood cells entering a constriction. *Biomicrofluidics* **2014**, *8*, 064123. [CrossRef] [PubMed]
41. Gabriele, S.; Benoliel, A.M.; Bongrand, P.; Theodoly, O. Microfluidic investigation reveals distinct roles for actin cytoskeleton and myosin II activity in capillary leukocyte trafficking. *Biophys. J.* **2009**, *96*, 4308–4318. [CrossRef] [PubMed]
42. Sharei, A.; Zoldan, J.; Adamo, A.; Sim, W.Y.; Cho, N.; Jackson, E.; Mao, S.; Schneider, S.; Han, M.-J.; Lytton-Jean, A.; *et al.* A vector-free microfluidic platform for intracellular delivery. *Proc. Natl. Acad. Sci. USA* **2013**, *110*, 2082–2087. [CrossRef] [PubMed]
43. Khan, Z.S.; Vanapalli, S.A. Probing the mechanical properties of brain cancer cells using a microfluidic cell squeezer device. *Biomicrofluidics* **2013**, *7*, 11806. [CrossRef] [PubMed]
44. Mak, M.; Erickson, D. A serial micropipette microfluidic device with applications to cancer cell repeated deformation studies. *Integr. Biol.* **2013**, *5*, 1374–1384. [CrossRef] [PubMed]
45. Lee, L.M.; Liu, A.P. A microfluidic pipette array for mechanophenotyping of cancer cells and mechanical gating of mechanosensitive channels. *Lab Chip* **2015**, *15*, 264–273. [CrossRef] [PubMed]
46. Guo, Q.; Park, S.; Ma, H. Microfluidic micropipette aspiration for measuring the deformability of single cells. *Lab Chip* **2012**, *12*, 2687–2695. [CrossRef] [PubMed]
47. Luo, Y.N.; Chen, D.Y.; Zhao, Y.; Wei, C.; Zhao, X.T.; Yue, W.T.; Long, R.; Wang, J.B.; Chen, J. A constriction channel based microfluidic system enabling continuous characterization of cellular instantaneous Young's modulus. *Sens. Actuat. B Chem.* **2014**, *202*, 1183–1189. [CrossRef]
48. Lange, J.R.; Steinwachs, J.; Kolb, T.; Lautscham, L.A.; Harder, I.; Whyte, G.; Fabry, B. Microconstriction arrays for high-throughput quantitative measurements of cell mechanical properties. *Biophys. J.* **2015**, *109*, 26–34. [CrossRef] [PubMed]
49. Preira, P.; Grandne, V.; Forel, J.M.; Gabriele, S.; Camara, M.; Theodoly, O. Passive circulating cell sorting by deformability using a microfluidic gradual filter. *Lab Chip* **2013**, *13*, 161–170. [CrossRef] [PubMed]
50. Mak, M.; Reinhart-King, C.A.; Erickson, D. Elucidating mechanical transition effects of invading cancer cells with a subnucleus-scaled microfluidic serial dimensional modulation device. *Lab Chip* **2013**, *13*, 340–348. [CrossRef] [PubMed]
51. Preira, P.; Valignat, M.-P.; Bico, J.; Theodoly, O. Single cell rheometry with a microfluidic constriction: Quantitative control of friction and fluid leaks between cell and channel walls. *Biomicrofluidics* **2013**, *7*, 024111. [CrossRef] [PubMed]
52. Tsai, C.H.; Sakuma, S.; Arai, F.; Kaneko, M. A new dimensionless index for evaluating cell stiffness-based deformability in microchannel. *IEEE Trans. Bio Med. Eng.* **2014**, *61*, 1187–1195. [CrossRef] [PubMed]
53. Zhao, Y.; Zhao, X.T.; Chen, D.Y.; Luo, Y.N.; Jiang, M.; Wei, C.; Long, R.; Yue, W.T.; Wang, J.B.; Chen, J. Tumor cell characterization and classification based on cellular specific membrane capacitance and cytoplasm conductivity. *Biosens. Bioelectron.* **2014**, *57*, 245–253. [CrossRef] [PubMed]
54. Zhao, Y.; Chen, D.; Luo, Y.; Li, H.; Deng, B.; Huang, S.-B.; Chiu, T.-K.; Wu, M.-H.; Long, R.; Hu, H.; *et al.* A microfluidic system for cell type classification based on cellular size-independent electrical properties. *Lab Chip* **2013**, *13*, 2272–2277. [CrossRef] [PubMed]
55. Zhao, Y.; Chen, D.; Li, H.; Luo, Y.; Deng, B.; Huang, S.B.; Chiu, T.K.; Wu, M.H.; Long, R.; Hu, H.; *et al.* A microfluidic system enabling continuous characterization of specific membrane capacitance and cytoplasm conductivity of single cells in suspension. *Biosens. Bioelectron.* **2013**, *43*, 304–307. [CrossRef] [PubMed]

56. Chen, J.; Xue, C.; Zhao, Y.; Chen, D.; Wu, M.H.; Wang, J. Microfluidic Impedance Flow Cytometry Enabling High-Throughput Single-Cell Electrical Property Characterization. *Int. J. Mol. Sci.* **2015**, *16*, 9804–9830. [CrossRef] [PubMed]

57. Wan, J.; Ristenpart, W.D.; Stone, H.A. Dynamics of shear-induced ATP release from red blood cells. *Proc. Natl. Acad. Sci. USA* **2008**, *105*, 16432–16437. [CrossRef] [PubMed]

58. Szeto, G.L.; Van Egeren, D.; Worku, H.; Sharei, A.; Alejandro, B.; Park, C.; Frew, K.; Brefo, M.; Mao, S.; Heimann, M.; *et al.* Microfluidic squeezing for intracellular antigen loading in polyclonal B-cells as cellular vaccines. *Sci. Rep.* **2015**, *5*, 10276. [CrossRef] [PubMed]

59. Huang, S.B.; Zhao, Z.; Chen, D.Y.; Lee, H.C.; Luo, Y.N.; Chiu, T.K.; Wang, J.B.; Chen, J.; Wu, M.H. A clogging-free microfluidic platform with an incorporated pneumatically-driven membrane-based active valve enabling specific membrane capacitance and cytoplasm conductivity characterization of single cells. *Sens. Actuat. B Chem.* **2014**, *190*, 928–936. [CrossRef]

60. Beattie, W.; Qin, X.; Wang, L.; Ma, H. Clog-free cell filtration using resettable cell traps. *Lab Chip* **2014**, *14*, 2657–2665. [CrossRef] [PubMed]

61. Zhao, Y.; Chen, D.; Luo, Y.; Chen, F.; Zhao, X.; Jiang, M.; Yue, W.; Long, R.; Wang, J.; Chen, J. Simultaneous characterization of instantaneous Young's modulus and specific membrane capacitance of single cells using a microfluidic system. *Sensors* **2015**, *15*, 2763–2773. [CrossRef] [PubMed]

62. Chen, J.; Zheng, Y.; Tan, Q.; Shojaei-Baghini, E.; Zhang, Y.L.; Li, J.; Prasad, P.; You, L.; Wu, X.Y.; Sun, Y. Classification of cell types using a microfluidic device for mechanical and electrical measurement on single cells. *Lab Chip* **2011**, *11*, 3174–3181. [CrossRef] [PubMed]

63. Zheng, Y.; Shojaei-Baghini, E.; Azad, A.; Wang, C.; Sun, Y. High-throughput biophysical measurement of human red blood cells. *Lab Chip* **2012**, *12*, 2560–2567. [CrossRef] [PubMed]

MDPI AG

St. Alban-Anlage 66

4052 Basel, Switzerland

Tel. +41 61 683 77 34

Fax +41 61 302 89 18

http://www.mdpi.com

Micromachines Editorial Office

E-mail: micromachines@mdpi.com

http://www.mdpi.com/journal/micromachines